SpringerBriefs in Statistics

For further volumes:
http://www.springer.com/series/8921

Kosto V. Mitov · Edward Omey

Renewal Processes

 Springer

Kosto V. Mitov
Aviation Faculty
National Military University "Vasil Levski"
Dolna Mitropolia
Bulgaria

Edward Omey
Faculty of Economics and Business
KU Leuven
Brussels
Belgium

ISSN 2191-544X   ISSN 2191-5458   (electronic)
ISBN 978-3-319-05854-2   ISBN 978-3-319-05855-9   (eBook)
DOI 10.1007/978-3-319-05855-9
Springer Cham Heidelberg New York Dordrecht London

Library of Congress Control Number: 2014935985

Mathematics Subject Classification (2010): 60K05, 60K10, 60K30, 26A12, 60-01

Printed on acid-free paper

Springer is part of Springer Science+Business Media (www.springer.com)

Preface

In this book we bring together an overview of renewal theory. We focus on *Renewal Processes* on the positive half-line and consider the continuous case, and we also deal with discrete renewal sequences. A prerequisite for these notes is an introduction to probability theory and stochastic processes. In an appendix we let the reader get acquainted with some of the important results of regularly varying functions.

The material presented in this book is intended for (graduate) students and researchers in mathematics, probability theory, and stochastic processes who need an introduction to renewal theory.

In Chap. 1 we give an overview of ordinary renewal theory on the positive half-line and we discuss the important renewal theoretic results such as the elementary renewal theorem and Blackwell's theorem. We also discuss a variety of rate of convergence theorems and obtain the limit distribution for the lifetime and residual lifetime of the renewal process. The chapter ends with a discussion of delayed renewal processes.

Chapter 2 is devoted to discrete renewal sequences and again we discuss the important theoretical results here. We give two proofs of the Erdös–Feller–Pollard theorem and again obtain several rate of convergence results here.

In Chap. 3 we discuss various extensions of the results so far. At first we give an overview of renewal theory for the infinite means case. Furthermore, we discuss alternating renewal processes, renewal reward processes, and the superposition of renewal processes. In this chapter, we also briefly discuss bivariate renewal theory.

In the Appendix, we give a short overview of some important definitions and properties of regularly varying functions.

The authors thank Evelyn Best who gave the idea for this book.

The authors also thank Prof. H. Walk and Prof. J. Dippon for their contribution to the present proof of Blackwell's renewal theorem, cf. Sect. 1.5.

They thank the reviewers for their valuable comments and suggestions which improved the book significantly and they also thank Veronika Rosteck who supported them in the preparation of the book.

Part of the book was written while E. Omey was visiting the Bulgarian Academy of Science and the department of Mathematics and Informatics of St. Kliment Ohridski University of Sofia. He thanks the department for its hospitality and support.

Pleven, Brussels   Kosto V. Mitov
December 2013   Edward Omey

Contents

Chapter 1
Renewal Processes

Abstract In this chapter, we give an overview of ordinary renewal theory on the positive half-line. We state the basic definitions and prove the important theorems on the renewal process and the renewal function. We give a detailed proof of the elementary renewal theorem and Blackwell's theorem and we also provide some rate of convergence results. Moreover, we discuss limit theorems for the lifetime processes and give a short overview of delayed renewal processes.

Keywords Ordinary renewal process · Delayed renewal process · Elementary renewal theorem · Blackwell's theorem · Key renewal theorem · Lifetime processes · Rate of convergence · Stationary renewal process

1.1 Definitions

Let on the probability space $(\Omega, \mathscr{A}, \Pr)$ be given a sequence

$$T, T_1, T_2, \ldots, T_n, \ldots$$

of nonnegative, independent identically distributed (i.i.d.) random variables (r.v.'s) with distribution function (d.f.) $F(x) = \Pr\{T \leq x\}$. To avoid trivialities, we shall always assume that $F(0) < 1$.

Definition 1.1 The sequence $\{S_n, n = 0, 1, 2, \ldots\}$ defined by

$$S_0 = 0, \quad S_n = T_1 + \cdots + T_n, \quad n = 1, 2, \ldots$$

is called a renewal sequence or a renewal process.

The quantities $S_0, S_1, S_2, \ldots$ are usually thought of as times of occurrence of some phenomena and are called *renewal events or renewal epochs or renewals*. The r.vs

K. V. Mitov and E. Omey, *Renewal Processes*, SpringerBriefs in Statistics,
DOI: 10.1007/978-3-319-05855-9_1, © The Author(s) 2014

T_n, $n = 1, 2, \ldots$ are the time intervals between the successive renewal events, and they are referred to as *interarrival times*.

Definition 1.2 The renewal counting process $\{N(t), t \geq 0\}$ associated with the renewal sequence $\{S_n, n \geq 0\}$ is defined by

$$N(t) = \sup\{n \geq 0 : S_n \leq t\} = \sum_{n=1}^{\infty} \mathbf{I}_{\{S_n \leq t\}},$$

where $\mathbf{I}_A$ is the indicator function of the event A, that is, $\mathbf{I}_A(\omega) = 1$, if $\omega \in A$, and $\mathbf{I}_A(\omega) = 0$, if $\omega \notin A$.

For convenience, one often uses a slight variant of this definition and one counts

$$M(t) = \inf\{n \geq 1 : S_n > t\}.$$

Since $M(t) = N(t) + 1$, the basic properties of $N(t)$ can be transferred to properties of $M(t)$ and vice versa. So, $\{N(t), t \geq 0\}$ has the obvious interpretation as the number of the renewal events in the interval $(0, t]$. If we consider $S_0 = 0$ as a renewal epoch, then the number of renewal events in the closed interval $[0, t]$ is equal to $M(t) = N(t) + 1$. In the book we will use both processes.

From Definition 1.2, we get the following three useful relations between $\{S_n\}$ and $\{N(t)\}$:

$$\{N(t) \leq n\} = \{S_{n+1} > t\}, \quad n \geq 0, \ t \geq 0, \tag{1.1}$$

$$S_{N(t)} \leq t < S_{N(t)+1}, \quad \text{almost surely for every } t \geq 0, \tag{1.2}$$

$$\{N(t) = n\} = \{S_n \leq t < S_{n+1}\}, \quad n \geq 0, \ t \geq 0. \tag{1.3}$$

Since the interarrival times T_n, $n = 1, 2, \ldots$ are nonnegative, their expectation

$$\mu := \mathbf{E}[T_n] = \int_0^{\infty} t \, dF(t) \in (0, \infty]$$

always exists, even though it may be infinite.

In some applications, we meet probability distributions, which assign a positive mass at ∞.

Definition 1.3 If

$$F(\infty) := \lim_{t \to \infty} F(t) = 1,$$

$F(t)$ is called *proper*, otherwise it is called *defective*.

If $F(\infty) < 1$, then T_n can be infinite with positive probability $1 - F(\infty)$. If this is the case, then the resulting renewal process is called *terminating*.

Fig. 1.1 Processes $A(t)$ and $B(t)$

Since the inequalities $S_N(t) \leq t$ and $S_{N(t)+1} > t$ are fulfilled with probability 1 the following three *lifetime processes* are well-defined.

Definition 1.4 (a) The process $A(t) = t - S_{N(t)}, t \geq 0$, is called the *spent lifetime (or age, forward recurrence time)*. It is the time elapsed from the last renewal epoch up to the moment of observation t;

(b) The process $B(t) = S_{N(t)+1} - t, t \geq 0$, is called the *residual lifetime (or excess, backward recurrence time)*. It represents the time elapsed from the moment of observation t up to the next renewal epoch;

(c) The process $C(t) = A(t) + B(t) = T_{N(t)+1}, t \geq 0$, is called the *lifetime (or spread, recurrence time)*. It is the time interval between the two successive renewals which contains the time of observation t (see Fig. 1.1).

1.2 Basic Properties

1.2.1 Properties of the Sample Paths

The renewal process $\{S_n, n = 0, 1, 2 \ldots\}$ is simply a sum of i.i.d. random variables. On the other hand, since the summands T_n are nonnegative, the sample paths of the sequence are nondecreasing, piecewise constant, and right-continuous. The counting process $\{N(t), t \geq 0\}$ plays a role of the inverse process to $\{S_n, n = 0, 1, 2, \ldots\}$. It is a continuous time process with piecewise constant right-continuous trajectories. The relations (1.1)–(1.3) allow to obtain the limit properties of $\{N(t), t \geq 0\}$ from the behavior of $\{S_n, n = 0, 1, 2, \ldots\}$.

Theorem 1.1 (i) *If $F(t)$ is a proper distribution function, then the trajectories of $\{N(t), t \geq 0\}$ are increasing to infinity with probability 1. Moreover with probability 1*

$$\lim_{t \to \infty} \frac{N(t)}{t} = \begin{cases} \mu^{-1}, & \text{if } \mu < \infty, \\ 0, & \text{if } \mu = \infty. \end{cases}$$

(ii) *If $q = F(\infty) < 1$, then there exists a last renewal epoch, i.e., $N(\infty) := \lim_{t \to \infty} N(t)$ is finite with probability 1. Moreover, for $n \geq 0$ and $x \geq 0$, the following relation holds*

$$\Pr\{N(\infty) = n, S_n \le x\} = (1 - F(\infty))F^{n*}(x).$$

As usual $F^{n*}(.)$ denotes the n-fold convolution of the d.f. $F(.)$ with itself (cf. Appendix A).

Proof (i) Denote $A = \{\omega : S_n(\omega)/n \to \mu, n \to \infty\}$. The Strong Law of Large Numbers yields $\Pr\{A\} = 1$. For any $\omega \in A$ there exists a constant $C > 0$ such that

$$n(\mu - C) \le S_n(\omega) \le n(\mu + C) \text{ for every } n = 0, 1, 2, \ldots.$$

Therefore, the inequality $N(t, \omega) > n$, which, by (1.1) is equivalent to $S_n(\omega) < t$, is fulfilled for every $t > n(\mu + C)$. Hence,

$$N(t, \omega) \to \infty, \quad t \to \infty.$$

From the last relation it follows that with probability 1,

$$\frac{S_{N(t)}}{N(t)} \to \mu \text{ and } \frac{N(t)}{N(t) + 1} \to 1, \quad \text{as } t \to \infty. \tag{1.4}$$

Assume that $\mu < \infty$. Using (1.2) one gets that almost surely,

$$\frac{S_N(t)}{N(t)} \le \frac{t}{N(t)} < \frac{S_{N(t)+1}}{N(t) + 1} \frac{N(t) + 1}{N(t)}.$$

These inequalities and (1.4) imply

$$\frac{t}{N(t)} \to \mu, \quad \text{as } t \to \infty.$$

Let $\mu = \infty$. Define the r.v.'s

$$T_n^c = \mathbf{I}_{\{T_n \le c\}} T_n, \quad n = 1, 2, \ldots,$$

where $c > 0$ is fixed. Clearly, $\mu^c = \mathbf{E}[T_n^c] = \int_0^c t \, dF(t) < \infty$ and $\mu^c \uparrow \infty$, $c \to \infty$. Let $\varepsilon > 0$ by fixed. Then $c > 0$ can be chosen such that $1/\mu^c < \varepsilon$. The following inequalities hold almost surely:

$$T_n^c \le T_n, \quad n = 1, 2, \ldots.$$

Therefore, $S_n^c = \sum_{i=1}^n T_i^c \le S_n$, $n = 0, 1, 2, \ldots$ and $N^c(t) = \max\{n : S_n^c \le t\} \ge N(t)$, $t > 0$. Then

$$0 \le \frac{N(t)}{t} \le \frac{N^c(t)}{t}, \quad t \ge 0,$$

almost surely. From the proof of the first case, we have that for all t large enough

$$\frac{N^c(t)}{t} \leq \frac{1}{\mu^c} + \varepsilon < 2\varepsilon,$$

almost surely. Therefore for all t large enough one has $0 \leq N(t)/t \leq 2\varepsilon$, almost surely. Since $\varepsilon > 0$ was arbitrary this completes the proof of (i).

(ii) Since $q = F(\infty) < 1$ then the event

$$\{N(\infty) = n\} = \{\text{the last renewal epoch has number } n\}$$

occurs if and only if

$$\{T_1 < \infty, T_2 < \infty, \ldots, T_{n-1} < \infty, T_n = \infty\}.$$

The probability of this event equals $q(1 - q)^{n-1}$, by the independence of T_i, $i = 1, 2, \ldots, n$. Hence, $N(\infty)$ has geometric distribution. Further on, we have

$$\{N(\infty) = n, S_n \leq x\}$$
$$= \{\text{ the last renewal epoch has number } n \text{ and happens before time } x\}$$
$$= \{S_n \leq x, T_{n+1} = \infty\}.$$

Now, the proof follows from the independence of T_{n+1} and S_n. $\qquad\qquad \triangle$

1.2.2 Distributions of the Process $N(t)$

For any fixed t the random variable $N(t)$ is integer valued.

Theorem 1.2 *The one-dimensional distributions of the renewal process $\{N(t), t \geq 0\}$ are determined by*

$$\Pr\{N(t) = n\} = F^{n*}(t) - F^{(n+1)*}(t), \quad n = 0, 1, \ldots.$$

Proof From (1.1), it follows that the events $\{N(t) \geq n\}$ and $\{S_n \leq t\}$ are equivalent. So,
$$\Pr\{N(t) \geq n\} = \Pr\{S_n \leq t\} = F^{n*}(t).$$

Therefore,

$$\Pr\{N(t) = n\} = \Pr\{N(t) \geq n\} - \Pr\{N(t) \geq n + 1\} = F^{n*}(t) - F^{(n+1)*}(t). \quad \triangle$$

The distribution of $N(t)$ can be found in an explicit form only in very few cases. Thus, the approximation given in the following theorem is useful in many situations.

Theorem 1.3 *Suppose that $0 < \sigma^2 = Var[T] < \infty$, then for each $x \in \mathbf{R}$*

$$\lim_{t \to \infty} \Pr \left\{ \frac{N(t) - t/\mu}{\sigma \sqrt{\mu^3 t}} \leq x \right\} = \Phi(x),$$

where $\Phi(x)$ is the standard normal distribution function.

Proof Recall that $\{N(t) \leq x\} = \{S_{[x]} \geq t\}$. Define $n = [a\sqrt{t} + tx/\mu]$, where $[y]$ denotes the integer part of y. As $t \to \infty$, we have $n \sim tx/\mu$ and

$$\frac{tx - \mu}{\sigma \sqrt{n}} \to -a \frac{\mu^{3/2}}{\sigma \sqrt{x}}.$$

Now we have

$$\left\{ \frac{N(t) - tx/\mu}{\sqrt{t}} \leq a \right\} = \{S_n \geq tx\}.$$

After normalizing, we find that

$$\Pr \{S_n \geq tx\} = \Pr \left\{ \frac{S_n - n\mu}{\sigma \sqrt{n}} \geq \frac{tx - n\mu}{\sigma \sqrt{n}} \right\} \to \Pr \left\{ Z \geq -a \frac{\mu^{3/2}}{\sigma \sqrt{x}} \right\}.$$

It follows that

$$\Pr \left\{ \frac{N(t) - tx/\mu}{\sqrt{t}} \leq a \right\} \to \Pr \left\{ Z \leq a \frac{\mu^{3/2}}{\sigma \sqrt{x}} \right\}$$

and the result follows. $\triangle$

We will formulate without proof a local limit theorem. Denote by $\phi(\xi) = \mathbf{E}[e^{i\xi T}]$ the characteristic function of T.

Theorem 1.4 (Omey and Vesilo [10]) *Suppose that $\sigma^2 = Var[T] < \infty$ and $|\phi(\xi)|^m$ is integrable for some $m \geq 1$. Then as $n \to \infty$,*

$$\sup_t \left| \frac{\sigma \sqrt{n}}{\mu} \Pr \{N(t) = n + 1\} - \Phi \left(\frac{t - \mu n}{\sigma \sqrt{n}} \right) \right| = o(1).$$

Moreover, if also $\mathbf{E}[T^3] < \infty$, then as $n \to \infty$,

$$\sqrt{n} \sup_t \left| \frac{\sigma \sqrt{n}}{\mu} \Pr \{N(t) = n + 1\} - \Phi \left(\frac{t - \mu n}{\sigma \sqrt{n}} \right) \right| = O(1).$$

The basic property of an ordinary renewal process is that the process starting from the first renewal epoch is an exact probabilistic copy of the whole process.

Theorem 1.5 (Basic finite-dimensional distribution property) *If $\{N(t), t \geq 0\}$ is an ordinary renewal process and $S_1 = T_1$ is the first renewal epoch, then the process $\{N(t + T_1) - 1, t \geq 0\}$ has the same finite-dimensional distributions as $\{N(t), t \geq 0\}$.*

Proof Let $0 < t_1 < t_2 < \ldots < t_n < \infty$ be an increasing sequence of times and $k_1, k_2, \ldots, k_n$, be a nondecreasing sequence of nonnegative integers. We will prove that

$$\Pr\{N(t_1 + T_1) - 1 = k_1, N(t_2 + T_1) - 1 = k_2, \ldots, N(t_n + T_1) - 1 = k_n\}$$
$$= \Pr\{N(t_1) = k_1, N(t_2) = k_2, \ldots, N(t_n) = k_n\}.$$

Transform the left-hand side as follows

$$\Pr\{N(t_1 + T_1) - 1 = k_1, N(t_2 + T_1) - 1 = k_2, \ldots, N(t_n + T_1) - 1 = k_n\}$$
$$= \Pr\{N(t_1 + T_1) = k_1 + 1, N(t_2 + T_1) = k_2 + 1, \ldots, N(t_n + T_1) = k_n + 1\}$$
$$= \Pr\{S_{k_1+1} \leq t_1 + T_1, S_{k_1+2} > t_1 + T_1, \ldots, S_{k_n+1} \leq t_n + T_1, S_{k_n+2} > t_n + T_1\}$$
$$= \Pr\{S_{k_1+1} - T_1 \leq t_1, S_{k_1+2} - T_1 > t_1, \ldots, S_{k_n+1} - T_1 \leq t_n, S_{k_n+2} - T_1 > t_n\}$$

Since

$$T_2 + T_3 + \cdots + T_{k_i+1} \text{ and } T_2 + T_3 + \cdots + T_{k_i+2}$$

have the same distributions as

$$T_1 + T_2 + \cdots + T_{k_i} \text{ and } T_1 + T_2 + \cdots + T_{k_i+1},$$

respectively, the chain of equalities continues as follows

$$= \Pr\{S_{k_1} \leq t_1, S_{k_1+1} > t_1, \ldots, S_{k_n} \leq t_n, S_{k_n+1} > t_n\}$$
$$= \Pr\{N(t_1) = k_1, N(t_2) = k_2, \ldots, N(t_n) = k_n\},$$

which completes the proof. △

The basic property is usually called "renewal argument". It allows to obtain a special type of integral equations for the quantities related to a given renewal process. The renewal argument will be often used in the following sections.

1.3 Renewal Function

1.3.1 Definition

Definition 1.5 The mean number of the renewal epochs

$$U(t) = \mathbf{E}[M(t)] = \mathbf{E}[N(t)] + 1$$

in the closed interval $[0, t]$ is called the *renewal function*.

From the definition of $N(t)$ one gets

$$U(t) = \mathbf{E}\left[\sum_{n=0}^{\infty} \mathbf{I}_{\{S_n \leq t\}}\right] = \sum_{n=0}^{\infty} \Pr\{S_n \leq t\} = \sum_{n=0}^{\infty} F^{n*}(t). \qquad (1.5)$$

We apply the renewal argument in order to obtain an integral equation for the renewal function.

Let t be a fixed time of observation. Conditionally, on the first renewal event $S_1 = T_1$ we have:

1. If $T_1 > t$, which happens with probability $1 - F(t)$, then $N(t) = 0$ and $\mathbf{E}[N(t)] = 0$.

2. If $T_1 = u \leq t$, which happens with probability $dF(u)$, then $N(t) = N(t-u)+1$ and $U(t) - 1 = \mathbf{E}[N(t)] = \mathbf{E}[N(t - u) + 1] = U(t - u)$.

Applying the formula for the total mathematical expectation we get for $t \geq 0$

$$U(t) - 1 = 0.(1 - F(t)) + \int_0^t U(t - u)dF(u),$$

i.e.,

$$U(t) = I(t) + \int_0^t U(t - u)dF(u), \qquad (1.6)$$

where

$$I(t) = \begin{cases} 0, & \text{if } t < 0 \\ 1, & \text{if } t \geq 0. \end{cases}$$

Equation (1.6) is a particular case of the general renewal equation which is discussed in Sect. 1.6.

It is intuitively obvious that under certain conditions $U(t)$ must be finite for every $t \geq 0$. The following theorem states this result.

Theorem 1.6 *Suppose that $F(t)$ is not concentrated at 0, i.e., $F(0) < 1$. Then for every $t \geq 0$, $U(t) < \infty$.*

In fact we will prove that the process $M(t) = N(t) + 1$ has finite moments of any order. Let us consider an example first.

Example 1.1 Assume that $\Pr\{T = 1\} = p \in (0, 1)$ and $\Pr\{T = 0\} = 1-p$. Then S_n has a binomial distribution and the number of experiments needed to have at least k successes is given by

$$M(k) = \inf \{n \geq 1 : S_n > k\}.$$

Hence, $M(k)$ has a negative binomial distribution:

$$\Pr\{M(k) = n\} = \binom{n-1}{k} p^{k+1} q^{n-k-1}, \ k \geq 0, n \geq k+1.$$

It is well known that the probability generating function (p.g.f.) of $M(k)$ is given by

$$\mathbf{E}[z^{M(k)}] = \left(\frac{pz}{1-qz}\right)^{k+1}.$$

Then we have $\mathbf{E}[M(k)] = (k+1)/p$ and for $r \geq 1$, $\mathbf{E}[M^r(k)] < \infty$. Let us mention that for positive values of t with $k \leq t < k+1$, we have almost surely

$$M(k) \leq M(t) \leq M(k+1).$$

Proof of Theorem 1.6 In general, we will prove that for all $t \geq 0$ and $r \geq 1$, $\mathbf{E}[M^r(t)] < \infty$. Since $F(0) < 1$, there exists a number a such that $p = \Pr\{T > a\} > 0$. For $n \geq 1$, define T_n^a as follows:

$$T_n^a = 0, \ \text{if} \ T_n \leq a \ \text{and} \ T_n^a = a, \ \text{if} \ T_n > a,$$

and let $M^a(t)$ denote the corresponding renewal process. Since $S_n^a \leq S_n$, we have $M(t) \leq M^a(t)$ and $\mathbf{E}[M^r(t)] \leq \mathbf{E}[(M^a(t))^r]$. Now the result follows from Example 1.1. △

In the next result, we compare $M(t)$ and $M(t+y)$ for any $t > 0$, $y > 0$, taking i.i.d. copies T_i^* and $M^*(t)$ of the original sequence T_i and $M(t)$. The lemma below gives a formula for the probabilities $a(r) = \Pr\{M(t+y) - M(t) = r\}$.

Lemma 1.1 *For $y > 0$ we have $a(0) = \Pr\{B(t) > y\}$ and*

$$a(r) = \Pr\{M^*(y - B(t)) = r\}, r \geq 1.$$

Proof By conditioning on $M(t)$, we have

$$a(0) = \sum_{n=1}^{\infty} \Pr\{M(t+y) = n, M(t) = n\}$$

$$= \sum_{n=1}^{\infty} \Pr\{M(t) = n, B(t) > y\} = \Pr\{B(t) > y\}.$$

For $r \geq 1$ we have

$$a(r) = \sum_{n=1}^{\infty} \Pr\{M(t) = n, M(t+y) = n + r\}$$

$$= \sum_{n=1}^{\infty} \Pr\left\{M(t) = n, \sum_{i=1}^{n+r-1} T_i \leq t + y < \sum_{i=1}^{n+r} T_i\right\}$$

$$= \sum_{n=1}^{\infty} \Pr\left\{M(t) = n, \sum_{i=n+1}^{n+r-1} T_i \leq y - B(t) < \sum_{i=n+1}^{n+r} T_i\right\}.$$

Since $\{M(t) = n, B(t)\}$ is independent of $\{T_j, j \geq n+1\}$, we have

$$a(r) = \sum_{n=1}^{\infty} \Pr\left\{M(t) = n, \sum_{i=1}^{r-1} T_i^* \leq y - B(t) < \sum_{i=1}^{r} T_i^*\right\}$$

$$= \sum_{n=1}^{\infty} \Pr\{M(t) = n, M^*(y - B(t)) = r\}$$

$$= \Pr\{M^*(y - B(t)) = r\}.$$

The result follows.   △

The mean number of renewal events in a half-open interval $(a, b]$ is equal to $U(b) - U(a)$. In this way, $U(t)$ defines a measure on $\mathbf{R}^+$ with an atom $U(0) = 1$ at the origin. The following result is useful.

Theorem 1.7 *For any $x > 0, y > 0$ the following inequalities hold true:*

$$0 \leq U(x + y) - U(x) \leq U(y).$$

Proof From Lemma 1.1, we have

$$U(x + y) - U(x) = E[M(x + y) - M(x)] = E[M^*(y - B(x))].$$

Since $B(x) > 0$, we obtain that

$$U(x + y) - U(x) \leq E[M^*(y)] = U(y).   △$$

Let us mention that the assertion of the theorem can be rewritten as $0 \leq U(x+y) \leq U(x) + U(y)$, which means that the renewal function is subadditive.

From (1.5) one has

$$\hat{U}(s) = \sum_{n=0}^{\infty} (\hat{F}(s))^n = \frac{1}{1 - \hat{F}(s)}, \quad s \geq 0, \qquad (1.7)$$

where $\hat{U}(s)$ and $\hat{F}(s)$ are the Laplace transforms of $U(.)$ and $F(.)$ (see Appendix A).

The following theorem gives the higher moments of $N(t)$ in terms of $U(t)$.

Theorem 1.8 *For $k \geq 1$ we have*

$$\mathbf{E}\left[\binom{N(t)+k}{k}\right] = U^{k*}(t).$$

Proof For $k = 1$, we obtain $\mathbf{E}[N(t) + 1] = U(t)$. In general we have

$$U_k(t) = \mathbf{E}\left[\binom{N(t)+k}{k}\right] = \sum_{n=0}^{\infty}\binom{n+k}{k}(F^{n*}(t) - F^{(n+1)*}(t))$$

$$= \sum_{n=0}^{\infty}\binom{n+k}{k}F^{n*}(t) - \sum_{n=0}^{\infty}\binom{n+k}{k}F^{(n+1)*}(t)$$

$$= 1 + \sum_{n=0}^{\infty}\left[\binom{n+k+1}{k} - \binom{n+k}{k}\right]F^{(n+1)*}(t)$$

$$= \sum_{n=0}^{\infty}\binom{n+k-1}{n}F^{n*}(t).$$

For the Laplace transform we find that

$$\hat{U}_k(s) = \sum_{n=0}^{\infty}\binom{n+k-1}{n}(\hat{F}(s))^n = \hat{U}(s)^k,$$

and the result follows. $\triangle$

If the distribution function $F(t)$ is defective then the measure U is finite. In this case, it is easy to be seen that

$$U(\infty) = \lim_{t\to\infty} U(t) = \frac{1}{1 - F(\infty)},$$

and $(1 - F(\infty))U(t)$ is a distribution function on $\mathbf{R}^+$.

1.3.2 Poisson Process

Suppose that the interarrival times $T_1, T_2, \ldots, T_n, \ldots$ are exponentially distributed with parameter $\lambda > 0$, that is, $F(x) = 1 - e^{-\lambda x}, x \geq 0; = 0, x < 0$. In this case, the renewal process $N(t), t \geq 0$ is a Poisson process, i.e., for any fixed $t \geq 0$,

$$\Pr\{N(t) = n\} = \frac{(\lambda t)^n}{n!}e^{-\lambda t}, \quad n = 0, 1, 2, \ldots,$$

and the mean number of renewal events in the interval $(0, t]$ is $\mathbf{E}[N(t)] = \lambda t,\ t \geq 0$. The mean number of renewals per time unit $\mathbf{E}[N(t)/t] = \lambda$ is called the intensity of the Poisson process.

The "lack of memory" property of the exponential distribution, $\Pr\{T_i > x + y | T_i > x\} = \Pr\{T_i > y\}$, provides that the Poisson process has a stronger property in its finite-dimensional distributions than that given in Theorem 1.5.

Let us denote by $N(t_0, t_0 + t],\ t \geq 0$, the number of renewal events in the time interval $(t_0, t_0 + t]$ for an arbitrary $t_0 \in [0, \infty)$. Then this process is also a Poisson process and its finite-dimensional distributions coincide with the finite-dimensional distributions of the Poisson process $N(t) := N(0, t),\ t \geq 0$. Let us assume that the last renewal event before time t_0 has the number $n,$, i.e., the following event occurs:

$$\{N(t_0) = n, S_n = \tau \leq t_0\} = \{S_n = \tau \leq t_0, T_{n+1} > t_0 - \tau\}.$$

Denote by Z_1 the residual lifetime between t_0 and the first renewal epoch after t_0. Then we have

$$\{Z_1 > x, N(t_0) = n, S_n = \tau \leq t_0\} = \{T_{n+1} > t_0 + x - \tau, N(t_0) = n, S_n = \tau \leq t_0\}.$$

Therefore,

$$
\begin{aligned}
\Pr\{Z_1 > x | N(t_0) &= n, S_n = \tau\} = \Pr\{T_{n+1} > t_0 + x - \tau | N(t_0) = n, S_n = \tau\} \\
&= \frac{\Pr\{T_{n+1} > t_0 + x - \tau, N(t_0) = n, S_n = \tau\}}{\Pr\{N(t_0) = n, S_n = \tau\}} \\
&= \frac{\Pr\{T_{n+1} > t_0 + x - \tau, S_n = \tau, T_{n+1} > t_0 - \tau\}}{\Pr\{S_n = \tau, T_{n+1} > t_0 - \tau\}} \\
&= \frac{\Pr\{T_{n+1} > t_0 + x - \tau, S_n = \tau\}}{\Pr\{S_n = \tau, T_{n+1} > t_0 - \tau\}} = \frac{\Pr\{T_{n+1} > t_0 + x - \tau\}\Pr\{S_n = \tau\}}{\Pr\{S_n = \tau\}\Pr\{T_{n+1} > t_0 - \tau\}} \\
&= \frac{\Pr\{T_{n+1} > t_0 + x - \tau\}}{\Pr\{T_{n+1} > t_0 - \tau\}} = \frac{e^{-\lambda(t_0 + x - \tau)}}{e^{-\lambda(t_0 - \tau)}} = e^{-\lambda x},
\end{aligned}
$$

the random variable Z_1 is exponentially distributed with parameter λ independently of t_0, n, and τ. So, Z_1 is independent and identically distributed with the interarrival times after that, which means that $N(t_0, t + t_0),\ t \geq t_0$ and $N(t),\ t \geq 0$ have the same finite-dimensional distributions.

Many characteristics of the Poisson process can be calculated explicitly. This fact will be used in the following sections for motivation and illustration of some results in the general case.

1.3.3   Other Examples

Apart from the Poisson process, there are only a few examples of renewal processes for which the renewal function can be expressed in a closed form. In this section, we present two examples in the continuous-time case. Three other examples are given in Chap. 2 and 3.

Example 1.2   Assume that T has a gamma distribution $\Gamma(2, \lambda)$. Its Laplace transform is given by

$$\widehat{F}(s) = \frac{\lambda^2}{(\lambda + s)^2}.$$

From (1.7) we find that

$$\widehat{U}(s) = \frac{1}{1 - \widehat{F}(s)} = \frac{(\lambda + s)^2}{(\lambda + s)^2 - \lambda^2}.$$

Therefore

$$\widehat{U}(s) = 1 + \lambda^2 \frac{1}{s(s + 2\lambda)} = 1 + \frac{\lambda}{2s} - \frac{1}{4}\frac{2\lambda}{2\lambda + s}.$$

Inverting, we conclude that

$$U(t) = 1 + \frac{\lambda}{2}t - \frac{1}{4}(1 - e^{-2\lambda t}), \quad t \geq 0.$$

Example 1.3   Let $T \sim U(0, 1)$. In this case we have

$$\widehat{F}(s) = \int_0^1 e^{-sx}dx = \frac{1 - e^{-s}}{s}.$$

It follows that

$$(\widehat{F}(s))^n = \frac{1}{s^n}(1 - e^{-s})^n = \frac{1}{s^n}\sum_{k=0}^{n}\binom{n}{k}(-1)^k e^{-sk}.$$

Now let $\delta_k(t) = 1$ if $t > k$ and $\delta_k(t) = 0$ if $t \leq k$ for $k = 0, 1, 2, \dots$. Also, define $G_n(t) := t^n/n!$ if $t \geq 0$; $G_n(t) := 0$ if $t < 0$. Clearly we have that

$$\widehat{\delta}_k(s) = \int_0^\infty e^{-st} d\delta_k(t) = e^{-sk}$$

$$\widehat{G}_n(s) = \frac{1}{(n-1)!} \int_0^\infty e^{-st} t^{n-1} dt = \frac{1}{s^n}.$$

It follows that

$$(\widehat{F}(s))^n = \sum_{k=0}^n \binom{n}{k} (-1)^k \widehat{G}_n(s) \widehat{\delta}_k(s).$$

As a consequence, we find that

$$F^{n*}(t) = \sum_{k=0}^n \binom{n}{k} (-1)^k \int_0^t \delta_k(t-y) dG_n(y).$$

For the integral, we find

$$\int_0^t \delta_k(t-y) dG_n(y) = \begin{cases} G_n(t-k), & \text{if } t \geq k \\ 0, & \text{if } t < k \end{cases}$$

or equivalently

$$\int_0^t \delta_k(t-y) dG_n(y) = \frac{1}{n!} ((t-k)^+)^n,$$

where $(t-u)^+ = \max(t-u, 0)$. We conclude that

$$F^{n*}(t) = \sum_{k=0}^n \frac{1}{n!} \binom{n}{k} (-1)^k ((t-k)^+)^n.$$

For another approach to get this formula, see (Feller [6], §I.9). For the renewal function $U(t)$, by taking sums, we obtain that

$$U(t) = \sum_{k=0}^\infty (-1)^k e^{(t-k)^+} \frac{1}{k!} ((t-k)^+)^k.$$

From a practical point of view, this formula is without much use. For small values of t, the formula can be useful; for $t \leq 1$, we get $U(t) = e^t$ and for $1 < t \leq 2$, we have $U(t) = e^t - e^{(t-1)}(t-1)$.

1.3.4 Elementary Renewal Theorem

We have already seen that $N(t)/t \to 1/\mu$, $t \to \infty$ almost surely. In case of exponentially distributed interarrival times the renewal function is $U(t) = \lambda t + 1$. In this case, $\mu = \mathbf{E}[T_i] = 1/\lambda$, hence

$$U(t)/t \to 1/\mu \text{ as } t \to \infty.$$

This relation holds true also in the general case. It is known as the *elementary renewal theorem*.

Theorem 1.9 (Elementary renewal theorem) *Suppose $F(0) < 1$ and $F(\infty) = 1$. Then*

$$\lim_{t \to \infty} \frac{U(t)}{t} = \begin{cases} \mu^{-1}, & \text{if } \mu < \infty, \\ 0, & \text{if } \mu = \infty. \end{cases}$$

Definition 1.6 Assume that on the common probability space are given a sequence $X_1, X_2, \ldots, X_n, \ldots$ of i.i.d. r.v.'s and an integer valued r.v. N such that for every $n = 1, 2, 3, \ldots$, the event $\{N \geq n\}$ may depend on the r.v.'s $X_1, X_2, \ldots, X_{n-1}$, but it is independent on the r.v.'s $X_n, X_{n+1}, \ldots$. The random variable N is called a *stopping time* for the sequence $X_n, n = 0, 1, 2 \ldots$.

In order to prove the theorem, we need the following lemma.

Lemma 1.2 (Wald's identity) *Assume that on the common probability space are given a sequence $X_1, X_2, \ldots, X_n, \ldots$ of i.i.d. r.v.'s and an integer valued r.v. N which is a stopping time for the sequence. Then*

$$\mathbf{E}[X_1 + X_2 + \cdots + X_N] = \mathbf{E}[X_1].\mathbf{E}[N].$$

Proof Let us denote by $I_n = \mathbf{I}_{\{N \geq n\}}$ the indicator function of the event $\{N \geq n\}$ ($I_n(\omega) = 1$, if $N(\omega) \geq n$ and $I_n(\omega) = 0$, if $N(\omega) < n$). The following equalities hold true:

$$\mathbf{E}[I_n] = \Pr\{N \geq n\}, \quad \mathbf{E}[N] = \sum_{n=1}^{\infty} \Pr\{N \geq n\} = \sum_{n=1}^{\infty} \mathbf{E}[I_n].$$

Since I_n does not depend on $X_n, X_{n+1}, \ldots$, it follows that for every n,

$$\mathbf{E}[I_n X_n] = \mathbf{E}[X_n].\Pr\{N \geq n\}.$$

On the other hand

$$S_N = \sum_{n=1}^{N} X_n = \sum_{n=1}^{\infty} I_n X_n.$$

Therefore,

$$\mathbf{E}[S_N] = \mathbf{E}\left[\sum_{n=1}^{\infty} I_n X_n\right] = \sum_{n=1}^{\infty} \mathbf{E}[I_n].\mathbf{E}[X_n]$$

$$= \mathbf{E}[X_1]\sum_{n=1}^{\infty} \mathbf{E}[I_n] = \mathbf{E}[X_1]\sum_{n=1}^{\infty} \Pr\{N \geq n\} = \mathbf{E}[X_1].\mathbf{E}[N]. \qquad \triangle$$

We present two proofs of Theorem 1.9.

Proof I of Theorem 1.9 The event $\{N(t) + 1 \geq n\}$ is equivalent to $\{N(t) \geq n - 1\}$, which in turn is equivalent to $\{T_1 + T_2 + \cdots + T_{n-1} \leq t\}$. So, for every $n = 1, 2, \ldots$ the event $\{N(t) + 1 \geq n\}$ does not depend on the r.v.'s $T_n, T_{n+1}, \ldots$. Hence, $M(t) = N(t) + 1$ is a stopping time for the sequence $T_n, n = 1, 2, \ldots$. Applying Wald's identity one gets

$$\mathbf{E}[S_{N(t)+1}] = \mathbf{E}[T].\mathbf{E}[N(t) + 1] = \mu U(t).$$

Since $S_{N(t)+1} > t$, almost surely, then $\mathbf{E}[S_{N(t)+1}] > t$. Therefore $\mu U(t) > t$, for $t > 0$ or

$$\frac{U(t)}{t} > \frac{1}{\mu}, \qquad \text{for } t > 0. \tag{1.8}$$

Let us note that $N(t)$ depends on $\{T_1 + T_2 + \cdots + T_n \leq t\}$. So, it is not a stopping time and it is not possible to apply the Wald's identity to $\mathbf{E}[S_{N(t)}]$ in order to obtain an upper limit.

Let $c > 0$ be a constant independent of n. Define the sequence of i.i.d. random variables

$$T_n^c = \begin{cases} T_n, & \text{if } T_n \leq c \\ c, & \text{if } T_n > c. \end{cases}$$

Define also the corresponding renewal sequence $\{S_n^c, n = 0, 1, 2\ldots\}$ and the renewal process $\{N^c(t), \ t \geq 0\}$. From the definition of T_n^c it follows that with probability 1,

$$T_n^c \leq T_n, \ \ S_n^c \leq S_n, \ \ \text{for every } n = 1, 2, \ldots.$$

Therefore with probability 1, $N^c(t) \geq N(t)$ for every $t \geq 0$ which leads to

$$U^c(t) - 1 = \mathbf{E}[N^c(t)] \geq \mathbf{E}[N(t)] = U(t) - 1$$

and (since $T_{n+1}^c \leq c$ with probability 1),

$$t < S_{N^c(t)+1}^c \leq t + c.$$

Now we have

$$t + c \geq \mathbf{E}[S_{N^c(t)+1}^c] = \mathbf{E}[T^c]\mathbf{E}[N^c(t) + 1] = \mathbf{E}[T^c]U^c(t) \geq \mathbf{E}[T^c]U(t).$$

Hence, using also (1.8), one gets

$$\frac{1}{\mu} < \frac{U(t)}{t} \leq \frac{1}{\mathbf{E}[T^c]} + \frac{c}{t\mathbf{E}[T^c]}. \tag{1.9}$$

Since $c > 0$ was arbitrary we set $c = \sqrt{t}$. Letting $t \to \infty$, one has $\mathbf{E}[T^c] \to \mathbf{E}[T] = \mu$, $t \to \infty$, and additionally $\frac{c}{t\mathbf{E}[T^c]} \to 0$, $t \to \infty$. Combining these relations and the inequalities (1.9) we complete the proof of the theorem when $\mu < \infty$. In case when $\mu = \infty$, one has to mention that also $\frac{1}{\mathbf{E}[T^c]} \to 0$, as $c \to \infty$. $\triangle$

Another proof is based on the Laplace transforms.

Proof II of Theorem 1.9 We have that for $s > 0$,

$$\hat{U}(s) = \frac{1}{1 - \hat{F}(s)}.$$

If the mean $\mu = \mathbf{E}[T] < \infty$, then

$$\lim_{s \to 0} \frac{1 - \hat{F}(s)}{s} = \mu,$$

and it follows that

$$\hat{U}(s) \sim \frac{1}{s\mu}, \quad \text{as } s \to 0.$$

From Karamata's theorem (see Appendix A, Theorem A.4) we get that

$$\frac{U(t)}{t} \to \frac{1}{\mu}, \quad \text{as } t \to \infty.$$

If $\mu = \infty$, we have $s\hat{U}(s) \to 0$ and $U(t)/t \to 0$. This completes the second proof.

                                                                                $\triangle$

1.4 Blackwell's Renewal Theorem

The elementary renewal theorem shows that the renewal function is asymptotically linear as $t \to \infty$. One might expect that $\frac{dU(t)}{dt}$ will be asymptotically equal to $1/\mu$, as $t \to \infty$. (In particular, this is true in case of Poisson process $\frac{dU(t)}{dt} = \lambda = \frac{1}{\mu}$ for all $t > 0$). Of course, in the general case the derivative of the renewal function does not always exist, but in any case one can consider the difference quotient

$$\frac{U(t) - U(t - h)}{h}$$

for any fixed $h > 0$ and $t > 0$ large enough. The limit behavior of this quotient was proved first by Erdös, Feller and Pollard (see Theorem 2.1) for the discrete time renewal process and then generalized by Blackwell in the continuous-time case.

In order to formulate and prove Blackwell's theorem, we need the following definition.

Definition 1.7 A distribution function $F(t)$ is said to be *lattice (or arithmetic)* with span $a > 0$ if the set of the points of increase of $F(t)$ is a subset of $\{x : x = na, n = 0, \pm 1, \pm 2, \ldots\}$ and a is the largest number satisfying this condition.

If $F(t)$ is not lattice it is called *nonlattice*. A special case of lattice distributions are those of integer valued random variables.

Theorem 1.10 (Blackwell's renewal theorem) *Suppose that*

$$F(0) < 1, \quad F(\infty) = \lim_{t \to \infty} F(t) = 1, \tag{1.10}$$

and

$$\mu = \int_0^\infty t dF(t) \in (0, \infty). \tag{1.11}$$

If $F(t)$ is nonlattice, then for any fixed $h > 0$

$$\lim_{t \to \infty} [U(t) - U(t - h)] = \frac{h}{\mu}. \tag{1.12}$$

If $F(t)$ is lattice with span $a > 0$, then for any fixed integer $l > 0$

$$\lim_{t \to \infty} [U(t) - U(t - la)] = \frac{la}{\mu}. \tag{1.13}$$

Let us note that if $\mu = \infty$, then the theorem still holds but the right-hand sides of (1.12) (or (1.13)) must be interpreted as zero.

The proof of the theorem in the lattice case will be given in Chap. 2.

There are several proofs in the nonlattice case. In the next section, the analytic proof that was published by Dippon and Walk [5] in 2005 is given.

1.5 Proof of Blackwell's Renewal Theorem

We need the following lemmas for the proof.

Lemma 1.3 *If a and b are points of increase for the distribution functions F and G, then $a + b$ is a point of increase for $F * G$.*

Proof If X and Y are independent with distribution functions F and G, respectively, then

$$\Pr\{|X + Y - a - b| < \varepsilon\} \geq \Pr\{|X - a| < \varepsilon/2\} \Pr\{|Y - b| < \varepsilon/2\}.$$

The right-hand side is positive for every $\varepsilon > 0$ if a and b are points of increase, and so $a + b$ is again a point of increase for $F * G$. △

Lemma 1.4 *Let F be a nonlattice distribution function concentrated on $[0, \infty)$ but $F(0) < 1$. The set Σ formed by the points of increase of $F, F^{2*}, F^{3*}, \ldots$ is asymptotically dense at ∞ in the sense that for a given $\varepsilon > 0$ and sufficiently large x the interval $(x, x + \varepsilon)$ contains points of Σ.*

Proof Let $\varepsilon > 0$ be fixed. Since F is nonlattice it is possible to choose two points $0 < a < b$ in the set Σ such that $h := b - a < \varepsilon$. Let I_n denote the interval $na < x \leq nb$. For every $n \geq n_0 = \left[\frac{a}{b-a}\right] + 1$ this interval contains $(na, (n + 1)a]$ as proper subinterval. Therefore, every point $x > n_0 a$ belongs to exactly one interval $(na, (n + 1)a]$ and hence it belongs to at least one among the intervals $I_1, I_2, \ldots, I_n \ldots$. By Lemma 1.3 the $n + 1$ points $na + kh = (n - k)a + kb$, $k = 0, \ldots, n$ belong to Σ. These points divide I_n into n subintervals of length h. Thus, every point $x > n_0 a$ is at a distance $\leq h/2$ from a point of Σ. This completes the proof. △

Proof of Theorem 1.10 (Dippon and Walk [5]).

A. Proof in the special case $\mathrm{E}[T_n^3] < \infty$ consists of 4 steps.

1st step: Let $h = 1$ without loss of generality. For fixed $c \in (0, 1)$ define the functions

$$z(t) = \begin{cases} 0, & t \leq -c, \\ (t + c)/c, & t \in [-c, 0], \\ 1, & t \in [0, 1], \\ (1 + c - t)/c, & t \in [1, 1 + c], \\ 0, & t \geq 1 + c, \end{cases}$$

(a majorant function of the indicator $1_{[0,1)}(.)$ of the unit interval), and

$$w(t) = (U * z)(t) = \int_{-\infty}^{\infty} z(t - x) dU(x)$$

for every $t \in \mathbf{R}$. It is easily checked that for every t,

$$w(t) \geq (U * \mathbf{1}_{[0,1)})(t) = U(t) - U(t-1), \tag{1.14}$$

and

$$
\begin{aligned}
((I - F) * w)(t) \\
= ((I - F) * (U * z))(t) = ((I - F) * U) * z)(t) = (I * z)(t) = z(t),
\end{aligned}
\tag{1.15}
$$

where $I(t) = F^{0*}(t)$ is a distribution function with unit mass at zero (see (1.6)). The following relation also holds true for every $t \geq 0$:

$$\int_0^\infty w(t-u)(1-F(u))du = \int_{-c}^t z(u)du. \tag{1.16}$$

Indeed, starting with the left-hand side of Eq. (1.16) one obtains for a fixed $t \geq 0$

$$\int_0^\infty w(t-u)(1-F(u))du = \int_{-\infty}^\infty w(t-u)(I-F)(u)du.$$

Then, in view of the definition of $w(\cdot)$,

$$
\begin{aligned}
\int_{-\infty}^\infty & w(t-u)(I-F)(u)du \\
&= \int_{-\infty}^\infty \left(\int_{-\infty}^\infty z(t-u-v)dU(v) \right)(I-F)(u)du \\
&= \int_{-\infty}^\infty \left(\int_{-\infty}^\infty z(t-u-v)(I-F)(u)du \right) dU(v).
\end{aligned}
$$

Changing variables $x = u + v$ in the inner integral and applying Fubini's theorem one gets

$$
\begin{aligned}
\int_{-\infty}^\infty & \left(\int_{-\infty}^\infty z(t-x)(I-F)(x-v)dx \right) dU(v) \\
&= \int_{-\infty}^\infty z(t-x) \left(\int_{-\infty}^\infty (I-F)(x-v)dU(v) \right) dx. \tag{1.17}
\end{aligned}
$$

However, we clearly have

$$\int_{-\infty}^{\infty} (I - F)(x - v)dU(v) = I(x), \qquad x \in \mathbf{R},$$

and hence the right-hand side of (1.17) becomes

$$\int_{-\infty}^{\infty} z(t - x)I(x)dx = \int_{0}^{\infty} z(t - x)dx = \int_{-c}^{t} z(y)dy,$$

the last equation following from the fact that $z(y)$ vanishes for $y \leq -c$.

For any fixed positive h the function $U(x) - U(x - h) \leq U(h)$ (see Theorem 1.7). Since $z(t)$ is Lipschitz continuous with bounded support then for any $x, x' \in \mathbf{R}$ such that $|x - x'| < 1$,

$$|w(x) - w(x')| \leq \int_{-\infty}^{\infty} |z(x - t) - z(x' - t)|dU(t) = C_1|x - x'|$$

for some constant C_1. Therefore, w is also Lipschitz continuous on $\mathbf{R}$. Furthermore,

$$w(x) = \int_{-\infty}^{\infty} z(x - t)dU(t) = \int_{\max(0,x-1-c)}^{x+c} z(x - t)dU(t)$$

$$\leq \int_{\max(0,x-1-c)}^{x+c} dU(t) = U(x + c) - U(x + c - (1 + 2c)) < C_2 < \infty.$$

Therefore, $w(x)$ is bounded on $\mathbf{R}$. Let $\{\tau_n'\}$ be a sequence such that as $n \to \infty$,

$$\tau_n' \to \infty \quad \text{and} \quad w(\tau_n') \to \alpha := \limsup_{x \to \infty} w(x) < \infty.$$

The sequence of the functions $w(x+\tau_n')$, $n = 1, 2, \ldots$ is equibounded and equicontinuous on $\mathbf{R}$. Therefore, by the Arzela-Ascoli theorem (see e.g., Feller [6], p. 270) a subsequence τ_n, $n = 1, 2, \ldots$ and a bounded continuous function g exist such that $w(x + \tau_n) \to g(x)$, $n \to \infty$ uniformly on bounded intervals, where $0 \leq g(x) \leq g(0) = \alpha$ and g is even Lipschitz continuous, by Lipschitz continuity of w. By the dominated convergence theorem one has for every $x \in \mathbf{R}$,

$$\int_{-\infty}^{\infty} w(x + \tau_n - t)d(I - F)(t) \to \int_{-\infty}^{\infty} g(x - t)d(I - F)(t), \quad n \to \infty.$$

On the other hand, by (1.15), for every $x \in \mathbf{R}$,

$$\int_{-\infty}^{\infty} w(x + \tau_n - t)d(I - F)(t) = z(x + t_n) \to 0 \ n \to \infty.$$

Therefore, for every $x \in \mathbf{R}$,

$$\int_{-\infty}^{\infty} g(x - t)d(I - F)(t) = 0, \tag{1.18}$$

that is

$$\int_{-\infty}^{\infty} g(x - t)dF(t) = g(x). \tag{1.19}$$

2nd step. In this step, it will be shown that Eq. (1.18) has no other solutions than constant, that is,

$$g(x) = g(0) = \alpha \text{ for all } x \in \mathbf{R}. \tag{1.20}$$

(In fact this is a special version of the Choquet and Deny theorem (see [3])).

Because of $\int_0^\infty (1 - F(t))dt = \mathbf{E}[T_1] < \infty$, from (1.18) one obtains that for every $x \in \mathbf{R}$,

$$\int_{-\infty}^{\infty} g(x - t)(I(t) - F(t))dt = const < \infty. \tag{1.21}$$

Indeed, since $g(x)$ is a bounded continuous function there exists an indefinite integral $G(x) = \int g(x)dx$, that is, $G(x)$ is defined on $\mathbf{R}$ and $G'(x) = g(x)$, for every $x \in \mathbf{R}$. Integrating both sides of (1.18) with respect to $x \in [0, y]$ and using Fubini's theorem one gets

$$\int_0^y \left(\int_{-\infty}^{\infty} g(x - t)d(I - F)(t) \right) dx = 0$$

$$\int_{-\infty}^{\infty} \left(\int_0^y g(x - t)dx \right) d(I - F)(t) = 0$$

$$\int_{-\infty}^{\infty} (G(y - t) - G(0 - t)) d(I - F)(t) = 0.$$

Integration by parts on the left-hand side leads to

$$\int_{-\infty}^{\infty} (G(y-t) - G(0-t))\, d(I-F)(t)$$

$$= (G(y-t) - G(0-t))\,(I-F)(t)|_{-\infty}^{\infty}$$

$$- \int_{-\infty}^{\infty} (g(y-t) - g(0-t)(I-F)(t)\, dt = 0.$$

Since $g(x) \le \alpha$ for all x and $(I-F)(t) \to 0$, as $t \to \pm\infty$ we get

$$- \int_{-\infty}^{\infty} (g(y-t) - g(0-t)(I-F)(t)\, dt = 0.$$

Therefore for any real y

$$\int_{-\infty}^{\infty} g(y-t)(I-F)(t)\, dt = \int_{-\infty}^{\infty} g(0-t)(I-F)(t)\, dt = const < \infty,$$

which is equivalent to (1.21). Therefore, we have to prove the uniqueness of the solution $g(x) = const/\mathbf{E}[T_1]$ of Eq. (1.21). Assume that $g_1(x)$ and $g_2(x)$ are two solutions to Eq. (1.21) and denote by $m(x) = g_1(x) - g_2(x)$. Now it is enough to show that if for a continuous real valued function $m(.)$,

$$\int_{-\infty}^{\infty} m(x-t)(I-F)(t)\, dt = 0, \quad \text{for all } x \tag{1.22}$$

then $m(x) \equiv 0$. The proof repeats the proof of a Tauberian theorem due to Kac [7]. We repeat it here for convenience. Note that the function $K(t) = (I-F)(t)$ satisfies the following conditions:

$$\int_{-\infty}^{\infty} |K(t)|\, dt < \infty$$

[by the third moment condition],

$$\kappa(\xi) = \int_{-\infty}^{\infty} K(t)e^{i\xi t}\, dt \ne 0 \quad \text{for all } -\infty < \xi < \infty \tag{1.23}$$

[by the fact that both F and I are nonlattice],

$$\int\limits_{-\infty}^{\infty} |t^2 K(t)| dt < \infty$$

[by the third moment condition].

Let us denote by Ψ the class of functions which have a continuous second derivative and which vanish outside a bounded interval. Let $\phi(\xi) \in \Psi$ and set

$$\Phi(t) = \int\limits_{-\infty}^{\infty} \phi(\xi) e^{i\xi t} d\xi.$$

Clearly, $\Phi(t)$ has bounded support and hence $\int_{-\infty}^{\infty} |\Phi(t)| dt$, and

$$\int\limits_{-\infty}^{\infty} \int\limits_{-\infty}^{\infty} |\Phi(t)||K(t-y)||m(y)| dt dy < \infty.$$

Using (1.22) and Fubini's theorem one gets

$$0 = \int\limits_{-\infty}^{\infty} \Phi(t) \left(\int\limits_{-\infty}^{\infty} K(t-y)m(y) dy \right) dt \qquad (1.24)$$

$$= \int\limits_{-\infty}^{\infty} m(y) \left(\int\limits_{-\infty}^{\infty} K(t-y)\Phi(t) dt \right) dy.$$

Clearly,

$$\int\limits_{-\infty}^{\infty} K(t-y)\Phi(t) dt = \int\limits_{-\infty}^{\infty} \kappa(\xi)\phi(\xi) e^{i\xi y} d\xi.$$

Thus, for each function $\phi(.) \in \Psi$, we have

$$0 = \int\limits_{-\infty}^{\infty} \int\limits_{-\infty}^{\infty} m(y)\kappa(\xi)\phi(\xi) e^{i\xi y} d\xi dy. \qquad (1.25)$$

By (1.24) it follows that $\kappa(\xi)$ has continuous second derivative and by (1.23) we also have that $\kappa(\xi) \neq 0$ for all ξ. Therefore, the product $\phi_1(\xi) = \kappa(\xi)\phi(\xi)$ is a function in Ψ for every function $\phi(\xi) \in \Psi$. So, we can rewrite (1.25) as

$$0 = \int_{-\infty}^{\infty} \int_{-\infty}^{\infty} m(y)\phi_1(\xi)e^{i\xi y}d\xi\, dy, \tag{1.26}$$

for $\phi_1(\xi) \in \Psi$. For any $\phi(\xi) \in \Psi$, $\phi(\xi - \beta) \in \Psi$ for every $\beta \in \mathbf{R}$. Therefore, for an arbitrary β, we can rewrite (1.26) as

$$0 = \int_{-\infty}^{\infty} \int_{-\infty}^{\infty} m(y)\phi_1(\xi - \beta)e^{i\xi y}d\xi\, dy.$$

Changing variables in the last integral implies that

$$0 = \int_{-\infty}^{\infty} m(y) \left(\int_{-\infty}^{\infty} \phi_1(\xi)e^{i\xi y}d\xi \right) e^{i\beta y}dy$$

$$= \int_{-\infty}^{\infty} m(y)\Phi_1(y)e^{i\beta y}dy.$$

Thus for all β

$$\int_{-\infty}^{\infty} m(y)\Phi_1(y)e^{i\beta y}dy = 0.$$

By the uniqueness of the Fourier transform, one gets $m(y)\Phi_1(y) = 0$ for almost all $y \in \mathbf{R}$. Because of the fact that ϕ_1 has compact support, Φ_1 is an entire function, and can be chosen not to be identically zero. Then it can have at most denumerable number of zeros. This implies that $m(y) = 0$ for almost all y, which completes the proof.

3rd step. Using (1.16) and $w(x + \tau_n) \to g(x)$ as $n \to \infty$ (pointwise) one obtains, by Fatou's lemma,

$$\int_0^{\infty} g(x - t)(1 - F(t))dt \le \lim_{n\to\infty} \int_0^{\infty} w(x + \tau_n - t)(1 - F(t))dt$$

$$= \lim_{n\to\infty} \int_{-c}^{x+\tau_n} z(y)dy = \int_{-c}^{1+c} z(y)dy = 1 + c.$$

By (1.20) one has

$$\lim_{x\to-\infty} g(x) = g(0) = \alpha. \tag{1.27}$$

Therefore, $g(x-t) \to \alpha$ as $x \to -\infty$ for each $t \in \mathbf{R}$. Applying again Fatou's lemma we get

$$\alpha \int_0^\infty (1 - F(t))dt \leq \lim_{x \to -\infty} \inf \int_0^\infty g(x-t)(1 - F(t))dt \leq 1 + c.$$

Therefore $\alpha \leq (1 + c)/\mathbf{E}[T_1]$. Now the inequality (1.14) and the definition of α imply

$$\limsup_{t \to \infty}(U(t) - U(t - 1)) \leq \frac{1 + c}{\mathbf{E}[T_1]},$$

which, in view of the fact that $c \in (0, 1)$ was arbitrary, is equivalent to

$$\limsup_{t \to \infty}(U(t) - U(t - 1)) \leq \frac{1}{\mathbf{E}[T_1]}. \tag{1.28}$$

4th step. Analogously to steps 1–3, for the case $\mathbf{E}[T_1^3] < \infty$, instead of the majorant function z of $1_{[0,1)}$, using the corresponding minorant function, one obtains

$$\liminf_{t \to \infty}(U(t) - U(t - 1)) \geq \frac{1}{\mathbf{E}[T_1]}. \tag{1.29}$$

Finally, the inequalities (1.28) and (1.29) imply the assertion of the theorem

$$\lim_{t \to \infty}(U(t) - U(t - 1)) = \frac{1}{\mathbf{E}[T_1]}.$$

B. Proof of Theorem 1.10 in general case $\mathbf{E}[T_1] \leq \infty$. The proof differs from the proof in the special case only in the second step and the use only of (1.27) instead of (1.20) in the third step.

2nd step. The aim is to show (1.27). Starting by the relation (1.19) which does not depend on the third moment condition, we have $(F^{n*} * g)(t) = g(t)$ for $n = 0, 1, 2, \ldots$. Therefore, for each $t \in \mathbf{R}$,

$$g(t) = \left[\left(\sum_{n=0}^\infty \frac{F^{n*}}{2^n} \right) * g \right](t).$$

Noticing $\max_{x \in \mathbf{R}} g(x) = g(0)$ and the continuity of g, one obtains

$$g(-x) = g(0) \text{ for all } x \in supp\left(\sum_{n=0}^\infty \frac{F^{n*}}{2^n}\right) = supp\left(\sum_{n=0}^\infty F^{n*}\right), \tag{1.30}$$

where *supp* refers to the support of measures. Indeed, if we assume that there exist integer n and $x \in suppF^{n*}$ such that $g(0 - x) < g(0)$, then by the continuity of g it follows that there exists an $\varepsilon > 0$ such that $g(0-y) < g(0)$ for each $y \in [x-\varepsilon, x+\varepsilon]$. Therefore,

$$g(0) = (F^{n*} * g)(0)$$

$$= \int_{-\infty}^{x-\varepsilon} g(0-y)dF^{n*}(y) + \int_{x-\varepsilon}^{x+\varepsilon} g(0-y)dF^{n*}(y)$$

$$+ \int_{x+\varepsilon}^{-\infty} g(0-y)dF^{n*}(y)$$

$$< \int_{-\infty}^{x-\varepsilon} g(0)dF^{n*}(y) + \int_{x-\varepsilon}^{x+\varepsilon} g(0)dF^{n*}(y)$$

$$+ \int_{x+\varepsilon}^{-\infty} g(0)dF^{n*}(y) = g(0).$$

Choose a continuous strictly increasing function $q : [0, 1] \to [0, \infty)$ such that $q(0) = 0, q(1) \le 1$ and

$$\int_{-\infty}^{\infty} t^2 q(1 - F(t))dt < \infty,$$

and define the distribution function $\tilde{F}$ by $\tilde{F}(t) = 0$ for $t < 0$ and $1 - \tilde{F}(t) = q(1-F(t))$ for $t \ge 0$. Then the probability distributions of F and $\tilde{F}$ are equivalent, and

$$\int_{0}^{\infty} t^3 d\tilde{F}(t) < \infty.$$

The renewal measures $U(.)$ and $\tilde{U}(.)$ are also equivalent and have the same support. By case A of the theorem, for each $h > 0$ one has

$$\tilde{U}(a) - \tilde{U}(a - h) \to \frac{h}{\int_0^\infty t d\tilde{F}(t)} > 0, \quad \text{as } a \to \infty.$$

By Lemma 1.4, the common support of the renewal measures $U(.)$ and $\tilde{U}(.)$ is asymptotically dense, i.e.,

$$dist\,(x, suppU) \to 0 \quad \text{as } x \to \infty.$$

The last relation together with (1.30) and the uniform continuity of g yields (1.27).

Remark 1.1 There are several ways to prove the theorem of Blackwell. Analytic proofs which are based on the renewal equation and the Choque–Deny theorem can be found in Feller [6] or Assmusen [2]. Alsmeyer [1] gives a proof based on Fourier analysis. Lindvall [8] provides a short proof based on the coupling of the renewal process with an independent, stationary renewal process.

We provide a probabilistic proof of the renewal theorem for the discrete time renewal processes in Chap. 2.

1.6 Renewal Equation

1.6.1 Definitions

The *renewal equation* is an integral equation of the form

$$Z(t) = z(t) + \int_0^t Z(t-u)dF(u), \quad t \geq 0 \tag{1.31}$$

or in shorter notation

$$Z(t) = z(t) + F * Z(t), \quad t \geq 0, \tag{1.32}$$

where $z : \mathbf{R} \to \mathbf{R}$, $z(t) = 0$, for $t < 0$ and $F(t)$ is a distribution function on $[0, \infty)$. Equation (1.6) for the renewal function $U(t)$ is a particular case of (1.31). Moreover, the renewal function $U(t)$ generated by $F(t)$ is strongly related to the solution of the renewal equation. In fact it determines the solution of the renewal equation as it is seen from the next theorem.

Theorem 1.11 *Suppose that $z(t)$ is bounded on finite intervals and vanishes on $(-\infty, 0)$. Then the only solution to the renewal equation (1.31) (or (1.32)) in the class of functions bounded on finite intervals and vanishing on $(-\infty, 0)$ is*

$$Z(t) = U * z(t) = \int_0^t z(t-u)dU(u), \quad t \geq 0. \tag{1.33}$$

Proof First, we show that $Z(t) = U * z(t)$ is a solution. Using the associative property of the convolution we have

$z(t) + F * Z(t) = z(t) + F * (U * z)(t)$
$$= z(t) + (F * U) * z(t) = z(t) + (U - I) * z(t) = z(t) + Z(t) - z(t) = Z(t).$$

Assume that $Z_1(t)$ and $Z_2(t)$ are two solutions of the renewal equation satisfying the conditions of the theorem. Denote by $V(t) = Z_1(t) - Z_2(t)$. For any fixed $t > 0$ we have

$$V(t) = \int_0^t V(t - u)dF(u).$$

Hence for every $n = 1, 2, \ldots,$

$$V(t) = \int_0^t V(t - u)dF^{n*}(u).$$

Since $V(t)$ is bounded on every finite interval one gets

$$|V(t)| \le \int_0^t |V(t - u)|dF^{n*}(u) \le \sup_{u \in [0,t]} |V(u)|F^{n*}(t) \to 0, \quad n \to \infty. \qquad \triangle$$

1.6.2 Key Renewal Theorem

Let us turn now to another theorem which together with the Blackwell's theorem is the heart of the renewal theory. In fact this theorem is the first one in a class of theorems concerning the asymptotic behavior of a convolution of the type (1.33). The theorems of this class are commonly referred to as *key renewal theorems*. The theorem below was proved by Smith [12], and we refer to it as Smith's key renewal theorem.

Definition 1.8 The function $z : \mathbf{R} \to \mathbf{R}$, that vanishes on $t < 0$, is said to be directly Riemann integrable (dRi) if for any $h > 0$ the series

$$\underline{\sigma} = h \sum_{n=0}^{\infty} \inf_{nh \le x \le (n+1)h} z(x),$$

$$\overline{\sigma} = h \sum_{n=0}^{\infty} \sup_{nh \le x \le (n+1)h} z(x)$$

converge absolutely and

$$\overline{\sigma} - \underline{\sigma} \to 0 \quad \text{as } h \to 0.$$

Example 1.4 1. If $z : [0, \infty) \to [0, \infty)$ is nonincreasing and $\int_0^\infty z(t)dt < \infty$ converges in the sense of the improper Riemman integral, then z is dRi. Indeed, if

$h > 0$ then for the series $\underline{\sigma}_h = h \sum_{n=1}^{\infty} z(nh)$ and $\overline{\sigma}_h = h \sum_{n=0}^{\infty} z(nh)$ one gets that

$$\underline{\sigma}_h \leq \int_0^{\infty} z(t)dt \leq \overline{\sigma}_h.$$

Furthermore,

$$\overline{\sigma}_h - \underline{\sigma}_h = h.z(0).$$

Hence, the difference can be made sufficiently small as $h \to 0$ and simultaneously $\underline{\sigma}_h \uparrow \int_0^{\infty} z(t)dt$ and $\overline{\sigma}_h \downarrow \int_0^{\infty} z(t)dt$.

2. If $z(t) = z_1(t) - z_2(t)$ where z_1 and z_2 are dRi, then z is also dRi.

3. If $z(t) \neq 0$ for $0 < a \leq t \leq b < \infty$, $z(t) = 0$, for $t \notin [a, b]$ and z is bounded, then it is dRi.

4. If $0 \leq z(t) \leq z_1(t)$, $t \in [0, \infty)$ and $z_1(t)$ is dRi, then z is also dRi.

Theorem 1.12 (Smith's key renewal theorem) *Let $z(t)$ be directly Riemann integrable and $z(t) = 0$ for $t < 0$. Suppose that $F(t)$ satisfies (1.10) and (1.11). If $F(t)$ is nonlattice, then*

$$Z(t) = U * z(t) \to \frac{1}{\mu} \int_0^{\infty} z(t)dt, \text{ as } t \to \infty.$$

If $F(t)$ is lattice with span $a > 0$ then

$$Z(t + na) \to \frac{a}{\mu} \sum_{k=0}^{\infty} z(t + ka), \text{ as } n \to \infty.$$

Theorem 1.13 *Theorems 1.10 and 1.12 are equivalent.*

Proof (Theorem 1.10 $\Rightarrow$ Theorem 1.12). Assume that the d.f. F is nonlattice and $\mu < \infty$. In order to simplify the proof we assume that $z(t) \downarrow 0$, $t \to \infty$ and $\int_0^{\infty} z(t)dt < \infty$, i.e., $z(t)$ is nonincreasing and z is integrable in the improper sense.

Let $h > 0$ and $0 < \varepsilon < 1$ be fixed. Then

$$\int_0^t z(t - u)dU(u) = \int_0^{t\varepsilon} + \int_{t\varepsilon}^t = J_1(t) + J_2(t).$$

We will estimate $J_2(t)$ first. For t large enough (e.g., $t(1 - \varepsilon) > 100\,h$) denote $n_h = \left[\frac{t(1-\varepsilon)}{h} \right]$ and subdivide the interval $[t\varepsilon, t]$ into subintervals of length h from the right to the left, as follows:

$$[t\varepsilon, t - n_h h], \ldots [t - 2h, t - h], [t - h, t].$$

It is not difficult to check that

$$t\varepsilon < t - n_h h < t\varepsilon + h, \tag{1.34}$$

that is, the leftmost subinterval has length less than h. Then

$$J_2(t) = \int_{t\varepsilon}^{t} z(t - u)dU(u)$$

$$= \int_{t\varepsilon}^{t-n_h h} z(t - u)dU(u) + \sum_{j=1}^{n_h} \int_{t-jh}^{t-(j-1)h} z(t - u)dU(u)$$

[since $z(t)$ is nonincreasing],

$$\leq \int_{t\varepsilon}^{t-n_h h} z(t - u)dU(u) + \sum_{j=1}^{n_h} z((j-1)h)\left(U(t-(j-1)h) - U(t-jh)\right)$$

$$= \int_{t\varepsilon}^{t-n_h h} z(t - u)dU(u) + \sum_{j=1}^{n_h} z((j-1)h)\left(U(t-jh+h) - U(t-jh)\right).$$

Because of $h > 0$ and $\varepsilon > 0$ are fixed, it is possible to choose t large enough, so that by Theorem 1.10

$$U(t - jh + h) - U(t - jh) \in \left((1 - \varepsilon)\frac{h}{\mu}, (1 + \varepsilon)\frac{h}{\mu}\right),$$

for all $j = 1, 2, \ldots, n_h$. Notice that from (1.34) one gets for every $j = 1, 2, \ldots, n_h$,

$$t - jh \geq t - n_h h > t\varepsilon \to \infty.$$

Therefore,

$$J_2(t) \leq \int_{t\varepsilon}^{t-n_h h} z(t - u)dU(u) + \frac{1 + \varepsilon}{\mu} \sum_{j=1}^{n_h} z((j-1)h)h$$

$$\leq \int_{t\varepsilon}^{t-n_h h} z(t - u)dU(u) + \frac{1 + \varepsilon}{\mu} \sum_{j=1}^{\infty} z((j-1)h)h$$

$$= \int_{t\varepsilon}^{t-n_h h} z(t - u)dU(u) + \frac{1 + \varepsilon}{\mu}\overline{\sigma_h}. \tag{1.35}$$

Now we will prove that as $t \to \infty$,

$$\int\limits_{t\varepsilon}^{t-n_h h} z(t-u)dU(u) \to 0.$$

Using the inequality (1.34) and the monotony of $U(t)$, we have

$$\int\limits_{t\varepsilon}^{t-n_h h} z(t-u)dU(u) \le z(n_h h)(U(t-n_h h)-U(t\varepsilon))$$

$$\le z(n_h h)(U(t\varepsilon + h)-U(t\varepsilon)).$$

From $z(t) \downarrow 0$ and Theorem 1.10 we get

$$z(n_h h)(U(t\varepsilon + h)-U(t\varepsilon)) \to 0, \quad t \to \infty.$$

The last relation and the inequality (1.35) yield that

$$\limsup_{t\to\infty} J_2(t) \le \frac{1+\varepsilon}{\mu}\overline{\sigma}_h.$$

In the same way, we get the lower bound

$$\liminf_{t\to\infty} J_2(t) \ge \frac{1-\varepsilon}{\mu}\underline{\sigma}_h.$$

For the estimation of $J_1(t)$ one has

$$0 \le J_1(t) = \int\limits_0^{t\varepsilon} z(t-u)dU(u) \le z(t-t\varepsilon)U(t\varepsilon)) \le z(t(1-\varepsilon))U(t)$$

$$= \frac{1}{1-\varepsilon}[t(1-\varepsilon)z(t(1-\varepsilon))]\frac{U(t)}{t} \to 0.\frac{1}{\mu}=0, \quad t \to \infty.$$

This follows from the Elementary renewal theorem and the following fact: If $z(t) \downarrow 0$ and $\int_0^\infty z(t)dt < \infty$, in improper sense then $tz(t) \to 0$, $t \to \infty$. Therefore, we can write

$$\frac{1-\varepsilon}{\mu}\underline{\sigma}_h \le \liminf_{t\to\infty}(J_1(t)+J_2(t)) \le \limsup_{t\to\infty}(J_1(t)+J_2(t)) \le \frac{1+\varepsilon}{\mu}\overline{\sigma}_h.$$

On the other hand, for every $h \to 0$ one has $\underline{\sigma}_h \uparrow \int_0^\infty z(t)dt$ and $\overline{\sigma}_h \downarrow \int_0^\infty z(t)dt$.

Since $\varepsilon > 0$ was arbitrary we compete the proof.

(Theorem 1.12 $\Rightarrow$ Theorem 1.10). From the renewal equation (1.6) for $U(t)$ it is not difficult to see that for any fixed $h > 0$ and $t \geq 0$ large enough, the function $Z(t) = U(t) - U(t - h)$ satisfies the following renewal equation:

$$U(t) - U(t - h) = I(t) - I(t - h) + \int_0^t [U(t - u) - U(t - h - u)]dF(u).$$

Since $z(t) = I(t) - I(t - h) = 1$ for $t \in [0, h)$ and vanishes outside the interval $[0, h]$, it is dRi. Therefore, we can apply Theorem 1.12 to get

$$\lim_{t \to \infty} [U(t) - U(t - h)] = \frac{1}{\mu} \int_0^\infty [I(t) - I(t - h)]dt = \frac{h}{\mu},$$

which completes the proof. $\triangle$

1.7 Rate of Convergence in Renewal Theorems

The basic theorems established in the previous three sections represent limits of the mean number of renewals in the intervals $[0, t]$, $[t - h, t]$ and the convolution $(U * z)(t)$ as $t \to \infty$. In any limit theorem, the question about the rate of convergence is very important. Below, we give two results concerning the rate of convergence in the Elementary renewal theorem and in the Blackwell's theorem.

Let us consider first $U(t) - \dfrac{t}{\mu}$. Assume that $\mu < \infty$. Then the function

$$F_0(t) = \frac{1}{\mu} \int_0^t (1 - F(u))du, \quad t \geq 0,$$

is well-defined proper distribution function concentrated on $[0, \infty)$. From (1.7) and (A.1) we have

$$\hat{U}(s)\hat{F}_0(s) = \hat{U}(s)\frac{(1 - \hat{F}(s))}{\mu s} = \frac{1}{\mu s}.$$

Hence $\dfrac{t}{\mu} = (U * F_0)(t)$. Then

$$U(t) - \frac{t}{\mu} = \int_0^t (1 - F_0(t - u))dU(u) = (U * Q)(t),$$

where $Q(t) = 1 - F_0(t)$. Let $R(t)$ denote the integral

$$R(t) = \int_0^t (1 - F_0(y))dy.$$

Theorem 1.14 (i) *Suppose that F is nonlattice and that* $\mu_2 = \mathbf{E}[T^2] < \infty$. *Then*

$$U(t) - \frac{t}{\mu} \to \frac{R(\infty)}{\mu}, \ as \ t \to \infty,$$

where $R(\infty) = \int_0^\infty (1 - F_0(y))dy < \infty$.

(ii) *Suppose that F is nonlattice and that* $\mu_2 = \infty$. *Then*

$$U(t) - \frac{t}{\mu} \sim \frac{1}{\mu}R(t), \ as \ t \to \infty.$$

Proof (i) Since $1 - F_0(t)$ is dRi, the first result follows from the Key renewal theorem (Theorem 1.12).

(ii) We choose $0 < \varepsilon < 1/\mu$ and $t^\circ > 0$ in such a way that for $t \geq t^\circ$,

$$\frac{1}{\mu} - \varepsilon \leq U(t+1) - U(t) \leq \frac{1}{\mu} + \varepsilon.$$

Now we write

$$(U * Q)(t) = \int_0^{t^\circ} Q(t-y)dU(y) + \int_{t^\circ}^t Q(t-y)dU(y) = J_1(t) + J_2(t).$$

Since $Q(t)$ is nonincreasing, we have $J_1(t) \leq Q(t-t^\circ)U(t^\circ)$. Since $R(t) \geq t(1 - F_0(t)) = tQ(t)$, and since $R(t)$ is increasing, it follows that

$$\frac{J_1(t)}{R(t)} \leq \frac{J_1(t)}{R(t-t^\circ)} \leq U(t^\circ)\frac{Q(t-t^\circ)}{(t-t^\circ)Q(t-t^\circ)} \to 0, \ as \ t \to \infty.$$

Now consider $J_2(t)$ and assume that t° is an integer. Then

$$J_2(t) = \sum_{n=t^\circ}^{[t]-1} \int_n^{n+1} Q(t-y)dU(y) + \int_{[t]}^t Q(t-y)dU(y)$$

$$= J_{21}(t) + J_{22}(t),$$

where as usual $[t]$ is the greatest integer, less or equal to t. Using the inequalities

$$Q(t-n)(U(n+1) - U(n)) \leq \int_n^{n+1} Q(t-y)dU(y) \leq Q(t-n-1)(U(n+1) - U(n))$$

one obtains for $J_{21}(t)$ that

$$\left(\frac{1}{\mu} - \varepsilon\right) \sum_{n=t^\circ}^{[t]-1} Q(t-n) \leq J_{21}(t) \leq \left(\frac{1}{\mu} + \varepsilon\right) \sum_{n=t^\circ}^{[t]-1} Q(t-n-1).$$

Now consider the integral $R(t)$. We have

$$R(t) = \int_0^{t^\circ} Q(t-y)dy + \int_{t^\circ}^{[t]} Q(t-y)dy + \int_{[t]}^t Q(t-y)dy$$

$$= R_1 + R_2 + R_3.$$

Since $R(t) \to \infty$ and $Q(t) \leq 1$, then

$$\frac{R_1 + R_3}{R(t)} \leq \frac{t^\circ + (t - [t])}{R(t)} \to 0.$$

For R_2 one obtains

$$R_2 = \sum_{n=t^\circ}^{[t]-1} \int_n^{n+1} Q(t-y)dy$$

so that

$$\sum_{n=t^\circ}^{[t]-1} Q(t-n) \leq R_2 \leq \sum_{n=t^\circ}^{[t]-1} Q(t-n-1)$$

or

$$\sum_{n=t^\circ}^{[t]-1} Q(t-n) \leq R_2 \leq \sum_{n=t^\circ-1}^{[t]-2} Q(t-n).$$

We find that

$$0 \leq R_2 - \sum_{n=t^\circ}^{[t]-1} Q(t-n) \leq Q(t-t^\circ+1) - Q(t-[t]+1).$$

Dividing by $R(t)$ and noting the right-hand side then goes to 0 it follows that

$$R(t) \sim R_2 \sim \sum_{n=t^\circ}^{[t]-1} Q(t-n).$$

We conclude that

$$\left(\frac{1}{\mu} - \varepsilon\right) \le \liminf_{t\to\infty} \frac{J_{21}(t)}{R(t)} \le \limsup_{t\to\infty} \frac{J_{21}(t)}{R(t)} \le \left(\frac{1}{\mu} + \varepsilon\right).$$

Next we consider $J_{22}(t)$. Since $Q(t) \le 1$, then

$$J_{22}(t) \le U(t) - U([t])$$

and it follows that

$$\frac{J_{22}(t)}{R(t)} \to 0.$$

Combining the estimates gives

$$\left(\frac{1}{\mu} - \varepsilon\right) \le \liminf_{t\to\infty} \frac{(U * Q)(t)}{R(t)} \le \limsup_{t\to\infty} \frac{(U * Q)(t)}{R(t)} \le \left(\frac{1}{\mu} + \varepsilon\right).$$

Now let $\varepsilon \to 0$ to obtain the proof of the result.   △

Remark 1.2 The same kind of proof shows the following result. Suppose that F is nonlattice and that $Q(x)$ is nonincreasing with $\int_0^\infty Q(t)dt = \infty$. Then

$$(U * Q)(t) \sim \frac{1}{\mu} \int_0^x Q(t)dt, \quad \text{as } t \to \infty.$$

Corollary 1.1 *Suppose that F is nonlattice and $1 < \alpha < 2$. Then*

$$1 - F(t) \in RV(-\alpha) \Longleftrightarrow U(t) - \frac{t}{\mu} \in RV(2-\alpha).$$

Both statements imply that

$$U(t) - \frac{t}{\mu} \sim \frac{t(1 - F_0(t))}{\mu(2-\alpha)}.$$

Proof Theorem 1.14 implies that

$$U(t) - \frac{t}{\mu} \sim \frac{1}{\mu} R(t).$$

If $1 - F(t) \in RV(-\alpha)$, then $1 - F_0(t) \in RV(1 - \alpha)$ and $R(t) \sim t(1 - F_0(t))/(2 - \alpha)$ (see Appendix A, Theorem A.1). Conversely, if $U(t) - t/\mu \in RV(2 - \alpha)$ then the theorem shows that $R(t) \in RV(2 - \alpha)$. Since $R(t)$ has monotone derivative and since $1 < \alpha < 2$, we find that $1 - F(t) \in RV(-\alpha)$ (see Appendix A, Theorem A.2). △

Definition 1.9 A distribution function G on the positive half-line belongs to the class $\mathscr{S}$ of subexponential distributions if it satisfies

$$\frac{1 - G^{2*}(x)}{1 - G(x)} \to 2, \quad \text{as } x \to \infty.$$

Let us note that the class $\mathscr{S}$ contains all d.f. G for which $1 - G$ varies regularly at infinity. Ney [9] proved the following refinement of Blackwell's theorem:

Theorem 1.15 *Assume that* $1 - F(t) \le c(1 - F_0(t))$ *and* $F_0 \in \mathscr{S}$.
(i) *Suppose that* F *is a nonsingular distribution with finite mean* μ. *Then as* $t \to \infty$, *we have*

$$\left| U(t + y) - U(t) - \frac{y}{\mu} \right| = O(1 - F_0(t)),$$

locally uniformly in y.
(ii) *If* F *is lattice with span 1 then as* $n \to \infty$ *we have*

$$\left| U(n + 1) - U(n) - \frac{1}{\mu} \right| = O(1 - F_0(n)).$$

1.7.1 Higher Moments of N(t)

Earlier (cf. Theorem 1.8) we proved that

$$\mathbf{E}\binom{N(t) + k}{k} = U^{k*}(t).$$

If the mean $\mu = \mathbf{E}[T] < \infty$, we have $U(t) \sim t/\mu$ and $U(t + y) - U(t) \to y/\mu$ as $t \to \infty$. Now we prove the following result:

Theorem 1.16 *If $\mu < \infty$, then for each $k \geq 1$*

$$U^{k*}(t) \sim \frac{t^k}{\mu^k k!} \tag{1.36}$$

and

$$U^{k*}(t+y) - U^{k*}(t) \sim \frac{t^{k-1}y}{\mu^k(k-1)!} \tag{1.37}$$

Proof The result holds for $k = 1$. For $k \geq 1$ we have that $U^{(k+1)*}(t) = U^{k*} * U(t)$. Since $\int_0^t U^{k*}(z)dz \to \infty$, as $t \to \infty$, we can use Theorem 1.14 (ii) and Remark 1.2 to obtain that

$$U^{(k+1)*}(t) \sim \frac{1}{\mu} \int_0^t U^{k*}(z)dz.$$

Since $U(t) \sim t/\mu$, it follows by induction that

$$U^{k*}(t) \sim \frac{t^k}{\mu^k k!}.$$

Indeed, (cf. Theorem A.1),

$$U^{2*}(t) \sim \frac{1}{\mu} \int_0^t U(z)dz \sim \frac{1}{\mu}\frac{t^2}{2\mu}, \quad t \to \infty.$$

Assuming $U^{k*}(t) \sim \frac{t^k}{\mu^k k!}$ we have

$$U^{(k+1)*}(t) \sim \frac{1}{\mu} \int_0^t U^{k*}(z)dz \sim \frac{1}{\mu}\frac{t^{k+1}}{(k+1)!\mu^k}, \quad t \to \infty.$$

Now we consider $U^{k*}(t+y) - U^{k*}(t)$. We will prove that

$$U^{k*}(t+y) - U^{k*}(t) \sim \frac{y}{\mu}U^{(k-1)*}(t).$$

For $k = 1$, we have the result of Blackwell. We proceed by induction and for $k \geq 2$ we write

$$U^{k*}(t+y) - U^{k*}(t) = \int_0^{t+y} U(t+y-z)dU^{(k-1)*}(z) - \int_0^t U(t-z)dU^{(k-1)*}(z)$$
$$= J_1(t) + J_2(t),$$

where

$$J_1(t) = \int_0^t (U(t+y-z) - U(t-z))dU^{(k-1)*}(z),$$

$$J_2(t) = \int_t^{t+y} U(t+y-z)dU^{(k-1)*}(z).$$

Choose $0 < \varepsilon < y/\mu$ and t° so that

$$\left| U(t+y) - U(t) - \frac{y}{\mu} \right| \le \varepsilon, \text{ for } t > t^\circ.$$

For $J_1(t)$, we write

$$J_1(t) = \left(\int_0^{t-t^\circ} + \int_{t-t^\circ}^t \right) (U(t+y-z) - U(t-z))dU^{(k-1)*}(z) = J_{11}(t) + J_{12}(t).$$

For $J_{11}(t)$ we have

$$\left(\frac{y}{\mu} - \varepsilon \right) U^{(k-1)*}(t-t^\circ) \le J_{11}(t) \le \left(\frac{y}{\mu} + \varepsilon \right) U^{(k-1)*}(t-t^\circ).$$

Using (1.36) we get that

$$\frac{y}{\mu} - \varepsilon \le \liminf_{t\to\infty} \frac{J_{11}(t)}{U^{(k-1)*}(t)} \le \limsup_{t\to\infty} \frac{J_{11}(t)}{U^{(k-1)*}(t)} \le \frac{y}{\mu} + \varepsilon.$$

Since $U(.)$ is bounded on bounded intervals, we have

$$0 \le J_{12}(t) + J_2(t)$$
$$\le C_1(U^{(k-1)*}(t) - U^{(k-1)*}(t-t^\circ)) + C_2(U^{(k-1)*}(t+y) - U^{(k-1)*}(t)).$$

By the induction hypothesis and (1.36) we have $J_{12}(t) + J_2(t) = O(t^{k-2})$, as $t \to \infty$. Therefore,

$$\frac{y}{\mu} - \varepsilon \leq \liminf_{t\to\infty} \frac{U^{k*}(t+y) - U^{k*}(t)}{U^{(k-1)*}(t)} \leq \limsup_{t\to\infty} \frac{U^{k*}(t+y) - U^{k*}(t)}{U^{(k-1)*}(t)} \leq \frac{y}{\mu} + \varepsilon.$$

Since $\varepsilon > 0$ was arbitrary, it follows that

$$\frac{U^{k*}(t+y) - U^{k*}(t)}{U^{(k-1)*}(t)} \to \frac{y}{\mu},$$

which completes the proof of (1.37). $\triangle$

In a similar way, the following result holds:

Proposition 1.1 *Suppose that $U(t)$ and $V(t)$ are the renewal functions of two independent renewal processes generated by X and Y respectively with $\mu = \mathbf{E}[X] < \infty$ and $v = \mathbf{E}[Y] < \infty$. Then as $t \to \infty$,*

$$\frac{(U * V)(t)}{t^2} \to \frac{1}{\mu v},$$

and

$$\frac{(U * V)(t+y) - (U * V)(t)}{t} \to \frac{y}{\mu v}.$$

1.8 Limit Theorems for Lifetimes

1.8.1 Renewal Equations for the Lifetimes Distributions

In this section, we will derive the renewal equations for the one-dimensional distributions of the spent, residual, and total lifetimes, $A(t)$, $B(t)$, and $C(t)$ respectively.

Theorem 1.17 (i) *The distribution function* $\Pr\{A(t) \leq x\}$, $t \geq 0$, $x \geq 0$, *of the age satisfies the following renewal equation*

$$\Pr\{A(t) \leq x\} = (1 - F(t))1_{[0,x]}(t) + \int_0^t \Pr\{A(t-u) \leq x\} \, dF(u). \quad (1.38)$$

(ii) *The tail of the distribution function* $\Pr\{B(t) > x\}$, $t \geq 0$, $x \geq 0$, *of the residual lifetime satisfies the following renewal equation*

$$\Pr\{B(t) > x\} = 1 - F(t+x) + \int_0^t \Pr\{B(t-u) > x\} \, dF(u). \quad (1.39)$$

(iii) *The distribution function of the process $C(t)$, $\Pr\{C(t) \leq x\}$ satisfies the renewal equation*

$$\Pr\{C(t) \leq x\} = (F(x) - F(t))1_{[0,x]}(t) + \int_0^t \Pr\{C(t-u) \leq x\}\, dF(u).$$

$$(1.40)$$

The indicator function $1_{[0,x]}(t) = 1$, if $t \in [0, x]$; and $1_{[0,x]}(t) = 0$, if $t \notin [0, x]$.

Proof In order to prove the theorem we will use the basic finite-dimensional distribution property of the renewal processes (see Theorem 1.5).

(i) Let us consider the event $\{A(t) \leq x\}$, conditioning on the first renewal epoch:

1. If $\{S_1 > t\}$, which happens with probability $\Pr\{S_1 > t\} = 1 - F(t)$, then $A(t) = t - S_0 = t - 0 = t$ and in this case the event $\{A(t) \leq x\}$ is equivalent to $\{t \leq x\}$. The last event has probability $1_{[0,x]}(t)$.

Therefore,

$$\Pr\{A(t) \leq x, S_1 > t\} = \Pr\{t \in [0, x], S_1 > t\} = 1_{[0,x]}(t)\Pr\{S_1 > t\} = 1_{[0,x]}(t)(1 - F(t)).$$

2. If for some $u \in [0, t]$, $\{S_1 \in [u, u + du)\}$, which has a probability $d\Pr\{S_1 \leq u\} = dF(u)$, then the process after the first renewal is a probabilistic copy of the whole process.

Therefore $\{A(t) \leq x, S_1 = u \in [0, t]\} = \{A(t - u) \leq x, S_1 = u \in [0, t]\}$, and

$$\Pr\{A(t) \leq x, S_1 = u \in [0, t]\} = \Pr\{A(t - u)) \leq x\}\Pr\{S_1 = u \in [0, t]\}.$$

An application of the total probability formula gives

$$\Pr\{A(t) \leq x\} = \Pr\{A(t) \leq x, S_1 > t\} + \int_0^t \Pr\{A(t) \leq x, S_1 = u\}\, d\Pr\{S_1 \leq u\}$$

$$= 1_{[0,x]}(t)(1 - F(t)) + \int_0^t \Pr\{A(t - u) \leq x\}\, dF(u).$$

This completes the proof of (1.38).

(ii) In this case, we repeat the same steps.

1. If $\{S_1 > t\}$ then $\{B(t) = S_1 - t > x\} = \{S_1 > t + x\}$. Therefore

$$\{S_1 > t, B(t) > x\} = \{S_1 > t + x\}.$$

2. If the first renewal is $S_1 = u \in [0, t]$ then

$$\{B(t) > x, S_1 = u \in [0, t]\} = \{B(t - u) > x, S_1 = u \in [0, t]\}.$$

Applying again the total probability formula we complete the proof of the theorem.

(iii) The proof goes the same way. We have only to note that if $\{S_1 > t\} = \{T_1 > t\}$ then $\{C(t) \le x\} = \{T_1 \le x\}$. $\triangle$

1.8.2 Limit Theorems

Now we shall apply the Key renewal theorem (Theorem 1.12) to each of the three renewal equations to obtain the limiting distribution of the processes $A(t)$, $B(t)$, and $C(t)$ as $t \to \infty$.

Theorem 1.18 *Suppose that the mean interarrival time is finite $(0 < \mu < \infty)$ and $F(.)$ is nonlattice.*

(i) *For $x \ge 0$,*

$$\lim_{t \to \infty} \Pr\{A(t) \le x\} = F_0(t) = \frac{1}{\mu} \int_0^x (1 - F(u))du.$$

(ii) *For $x \ge 0$,*

$$\lim_{t \to \infty} \Pr\{B(t) \le x\} = F_0(t) = \frac{1}{\mu} \int_0^x (1 - F(u))du.$$

(iii) *For $x \ge 0$*

$$\lim_{t \to \infty} \Pr\{C(t) \le x\} = \frac{1}{\mu} \int_0^x u dF(u). \tag{1.41}$$

Proof Let $x \ge 0$ be fixed.

(i) The function $1_{[0,x]}(t)(1 - F(t)) \neq 0$ is bounded for $t \in [0, x]$ and vanishes outside the interval $[0, x]$. Hence it is directly Riemann integrable. Applying the Key renewal theorem to the Eq. (1.38), one obtains

$$\lim_{t\to\infty} \Pr\{A(t) \le x\} = \frac{1}{\mu} \int_0^\infty 1_{[0,x]}(u)(1-F(u))du$$

$$= \frac{1}{\mu} \int_0^x (1-F(u))du = F_0(x).$$

(ii) Since the function $1 - F(t + x)$ is monotone decreasing and $\int_0^\infty (1 - F(t + x))dt \le \int_0^\infty (1 - F(t))dt = \mu < \infty$, it is also directly Riemann integrable. Applying the Key renewal theorem to Eq. (1.39) we obtain

$$\lim_{t\to\infty} \Pr\{B(t) > x\} = \frac{1}{\mu} \int_0^\infty (1-F(u+x))du = \frac{1}{\mu} \int_x^\infty (1-F(u))du.$$

Therefore,

$$\lim_{t\to\infty} \Pr\{B(t) \le x\} = 1 - \frac{1}{\mu} \int_x^\infty (1-F(u))du$$

$$= \frac{1}{\mu} \int_0^\infty (1-F(u))du - \frac{1}{\mu} \int_x^\infty (1-F(u))du$$

$$= \frac{1}{\mu} \int_0^x (1-F(u))du = F_0(x).$$

(iii) The proof of (1.41) is similar. In this case, we apply the Key renewal theorem to Eq. (1.40), since the function $(F(x) - F(t))1_{[0,x]}(t)$ vanishes outside the interval $[0, x]$ it is directly Riemann integrable. So

$$\lim_{t\to\infty} \Pr\{C(t) \le x\} = \frac{1}{\mu} \int_0^x (F(x) - F(u))du = \frac{x}{\mu}F(x) - \frac{1}{\mu} \int_0^x F(u)du.$$

An integration by parts completes the proof of (1.41). △

Let us denote by T_0 the random variable with distribution function $F_0(x)$. If the interarrival times T_n have finite second moment then T_0 has finite mean

$$\mu_0 := \mathbf{E}[T_0] = \int_0^\infty u dF_0(u) = \int_0^\infty u \frac{1}{\mu}(1 - F(u)) du = \frac{1}{2\mu} \int_0^\infty (1 - F(u)) du^2$$

$$= \frac{1}{2\mu} \left((1 - F(u)) u^2 \Big|_0^\infty + \int_0^\infty u^2 dF(u) \right) = \frac{\mathbf{E}[T^2]}{2\mu}.$$

Theorem 1.19 *Assume that* $\mathbf{E}[T^2] < \infty$. *Then with probability 1 we have*

$$\lim_{t \to \infty} \frac{1}{t} \int_0^t A(u) du = \mu_0, \qquad \lim_{t \to \infty} \frac{1}{t} \int_0^t B(u) du = \mu_0.$$

Proof By considering the graph of $A(t)$ as a series of triangles positioned at times S_n, the following inequalities are fulfilled with probability 1:

$$\frac{1}{t} \sum_{i=1}^{N(t)} \frac{T_i^2}{2} \le \frac{1}{t} \int_0^t A(u) du \le \frac{1}{t} \sum_{i=1}^{N(t)+1} \frac{T_i^2}{2}$$

or

$$\frac{N(t)}{t} \frac{1}{N(t)} \sum_{i=1}^{N(t)} \frac{T_i^2}{2} \le \frac{1}{t} \int_0^t A(u) du \le \frac{N(t)+1}{t} \frac{1}{N(t)+1} \sum_{i=1}^{N(t)+1} \frac{T_i^2}{2}.$$

Since $N(t) \to \infty$ with probability 1 and $N(t)/t \to 1/\mu$ with probability 1, we have only to apply the SLLN to the sums

$$\frac{1}{N(t)} \sum_{i=1}^{N(t)} \frac{T_i^2}{2} \to \frac{\mathbf{E}[T^2]}{2} \quad \text{and} \quad \frac{1}{N(t)+1} \sum_{i=1}^{N(t)+1} \frac{T_i^2}{2} \to \frac{\mathbf{E}[T^2]}{2}$$

to complete the proof of the first limit. The proof of the second limit is similar. △

1.9 Delayed Renewal Processes

1.9.1 Definition

In this section, we consider a renewal sequence $S_0 = 0$ and $S_n = \sum_{i=1}^n T_i$ but we assume that the first interarrival time T_1 has a different distribution from the remaining interarrival times $T_i, i \ge 2$. The corresponding renewal counting process is defined as before:

$$N^d(t) = \max\{n \geq 0 : S_n \leq t\}.$$

Definition 1.10 The stochastic process $\{N^d(t), t \geq 0\}$ is called a delayed renewal process.

Note that a delayed renewal process can be easily obtained by an ordinary renewal process if the initial time of observation is not $t = 0$ but is chosen at random in the interval $(0, \infty)$. In this case $T_1 = B(t)$, where $B(t)$ is the residual lifetime.

1.9.2 The Main Renewal Theorems are Still True

For convenience, let $G(t) = \Pr\{T_1 \leq t\}$ and $F(t) = \Pr\{T_i \leq t\}, i \geq 2$. As before, we have

$$\Pr\{N^d(t) \leq n\} = \Pr\{S_{n+1} > t\}, \quad \Pr\{M^d(t) > n\} = \Pr\{S_n \leq t\}.$$

The delayed renewal function is defined by

$$U^d(t) = \mathbf{E}[N^d(t)] = \sum_{n=1}^{\infty} \Pr\{S_n \leq t\}.$$

Using $\Pr\{S_n \leq t\} = G * F^{(n-1)*}(t)$, we obtain that

$$U^d(t) = \sum_{n=1}^{\infty} G * F^{(n-1)*}(t) = (G * U)(t), t \geq 0. \qquad (1.42)$$

Using Laplace transforms, we also have $\widehat{U^d}(s) = \widehat{G}(s)\widehat{U}(s)$. The main results of ordinary renewal theory remain valid here.

Theorem 1.20 *Suppose that F and G are nonlattice and that $\mu = \mathbf{E}[T_2] < \infty$. Then as $t \to \infty$,*

(i) $\dfrac{U^d(t)}{t} \to \dfrac{1}{\mu}$;

(ii) $U^d(t+h) - U^d(t) \to \dfrac{h}{\mu}$;

(iii) *If F and z satisfy the conditions of Theorem 1.12, we have*

$$\lim_{t \to \infty} (U^d * z)(t) = \frac{1}{\mu} \int_0^{\infty} z(x)dx.$$

Proof (i) From the expression of $\widehat{U}^d(s)$, we obtain that $\widehat{U}^d(s) \sim s/\mu$ as $s \to 0$. Now we can use Karamata's theorem (Theorem A.4).

(ii) Choose t° so that

$$\frac{h}{\mu} - \varepsilon \leq U(t+h) - U(t) \leq \frac{h}{\mu} + \epsilon, t \geq t^\circ.$$

Now we write

$$U^d(t+h) - U^d(t) = J_1(t) + J_2(t) + J_3(t),$$

where

$$J_1(t) = \int_0^{t-t^\circ} (U(t+h-y) - U(t-y))dG(y),$$

$$J_2(t) = \int_{t-t^\circ}^{t} (U(t+h-y) - U(t-y))dG(y),$$

$$J_3(t) = \int_{t}^{t+h} (U(t+h-y) - U(t-y))dG(y).$$

Clearly, we have $(h/\mu - \varepsilon)G(t-t^\circ) \leq J_1(t) \leq (h/\mu + \varepsilon)G(t-t^\circ)$. Since $U(.)$ is bounded on bounded intervals, for some constant C we have

$$0 \leq J_2(t) + J_3(t) \leq C(G(t) - G(t-t^\circ) + G(t+h) - G(t)).$$

It follows that $\lim_{t\to\infty} J_2(t) + J_3(t) = 0$. Since $G(t-t^\circ), G(t+h) \to 1$, and ε was arbitrary, the result follows.

(iii) The proof is similar to the proof of Theorem 1.13.   △

1.9.3 Stationary Renewal Process

The mean number of the renewal events in the interval of fixed length h converges to a constant proportional of the interval length and inverse proportional of the time between the renewals independently of the distribution of the initial renewal time. For the delayed renewal processes this is true only under the condition that $G(t)$ is proper.

In Sect. 1.3.2 we saw that for the Poisson process $N(t) = N(0, t], t \geq 0$, the number of renewals $N(t_0, t_0+t], t \geq 0$ in the interval $(t_0, t_0+t]$ for an arbitrary $t_0 \in [0, \infty)$ defines another process, which is also a Poisson process and its finite-dimensional distributions coincide with the finite-dimensional distributions of $N(t), t \geq 0$.

Furthermore, the Renewal function $U(t) = t/\mu$ is linear and the mean number of renewal events in an interval $(t, t + h]$ of fixed length $h > 0$ is h/μ independent of t.

Are there other renewal processes that exhibit similar properties?

We will prove first that if the distribution function $F(t)$ of the random variables $T_n, n \geq 2$, is known, it is possible to choose the distribution function of T_1, $G(t)$ such that $U^d(t) = t/\mu$, for every $t \geq 0$.

From (1.42), we have that $U^d(t)$ is the solution of the renewal equation

$$U^d(t) = G(t) + \int_0^t U^d(t - u)dF(u), \quad t \geq 0.$$

Substituting $U^d(t) = t/\mu$ we get

$$G(t) = \frac{t}{\mu} - \frac{1}{\mu} \int_0^t (t - u)dF(u), \quad t \geq 0,$$

which shows that

$$G(t) = F_0(t) := \frac{1}{\mu} \int_0^t (1 - F(u))du, \quad t \geq 0.$$

Therefore, if the renewal function is linear $U^d(t) = \frac{t}{\mu}$ then the distribution function of T_1 must be $F_0(t)$.

Conversely, if the distribution function of T_1 is $F_0(t)$ then $U^d(t) = t/\mu$. Indeed, $F_0(t)$ has a density $(1 - F(t))/\mu$. Hence, the Laplace transform of $F_0(t)$ is

$$\hat{F}_0(s) = \int_0^\infty e^{-st} \frac{1 - F(t)}{\mu} dt = \frac{1}{\mu} \left(\int_0^\infty e^{-st} dt - \int_0^\infty e^{-st} F(t)dt \right)$$

$$= \frac{1}{\mu} \left(\frac{1}{s} - \frac{1}{s} \int_0^\infty e^{-st} dF(t) \right) = \frac{1 - \hat{F}(s)}{s\mu},$$

with $\hat{F}(s)$ denoting the Laplace transform of $F(t)$. On the other hand, in this case, we have $U^d(t) = (U * F_0)(t)$ for $t \geq 0$. Hence, the Laplace transform of $U^d(t)$ is given by

$$\hat{U}^d(s) = \frac{\hat{F}_0(s)}{1 - \hat{F}(s)} = \frac{1}{s\mu}$$

for all $s > 0$ and thus coincides with the Laplace transform of the measure dx/μ.
Since the Laplace transform uniquely determines a function on $[0, \infty)$ then $U^d(t) = t/\mu$, $t \geq 0$. Hence, $U^d(t) - U^d(t-h) = \dfrac{h}{\mu}$, for every fixed $h > 0$ and every $t \geq h$.

We have already seen (Theorem 1.18 (ii)) that the limiting distribution of the residual lifetime $B(t)$ is $F_0(t)$. Let us consider the delayed renewal process with distribution function $F_0(t)$ of T_1. We will prove that for all $t \geq 0$ the distribution of the residual lifetime is exactly $F_0(t)$, i.e., $\Pr\{B(t) \leq x\} = F_0(x)$ does not depend ot t.

Similar to the proof of Theorem 1.17 (ii) we have

$$\Pr\{B(t) > x\} = 1 - F_0(t+x) + \int_0^t (1 - F(t+x-u))dU^d(u).$$

Using the definition of $F_0(t)$ and substituting $U^d(t) = \frac{t}{\mu}$ we obtain

$$\Pr\{B(t) > x\} = 1 - \frac{1}{\mu}\int_0^{t+x}(1 - F(u))du + \frac{1}{\mu}\int_0^t(1 - F(t+x-u))du$$

$$= \frac{1}{\mu}\int_x^\infty (1 - F(u))du,$$

which completes the proof.

Now we will prove that the delayed renewal process $N^d(t), t \geq 0$, with $G(t) = F_0(t)$ has stationary increments.

For a fixed $t > 0$ and $s \geq 0$, $N_s^d(t) = N^d(s+t) - N^d(s)$ counts the renewal events in a sequence $S'_n, n = 1, 2, \ldots$ where $T'_1 = B(s)$ and the following inter-arrival times $T'_n, n \geq 2$ are independent of T'_1 and have distribution function $F(t)$. On the other hand, we have proved above that $B(s)$ has the same distribution function $F_0(t)$, as $B(0) = T_1$. Hence $N^d(t) = N^d(0, t]$ has the same distribution as $N_s^d(t) = N^d(t+s) - N(s) = N^d(s, s+t]$.

The delayed renewal process $N^d(t)$ with $G(t) = F_0(t)$ is called *stationary*.

Remark 1.3 More detailed and comprehensive treatment of the characterizations of stationary renewal processes involving the theory of point processes and Markov processes can be found in Daley and Ver-Jones [4], Ch.3,4 or in Serfozo [11], Ch. 2.

1.9.4 An Interesting Example

In the previous two sections, we discussed the properties of the delayed renewal processes. From (1.42) it follows that

$$\widehat{U}^d(s) = \widehat{G}(s)\widehat{U}(s) = \frac{\widehat{G}(s)}{1 - \widehat{F}(s)}.$$

In Sect. 1.9.3 we saw that the clever choice

$$G(x) = F_0(x) = \frac{1}{\mu}\int_0^x (1 - F(u))du$$

gives $\widehat{G}(s) = (1 - \widehat{F}(s))/(\mu s)$ and as a consequence we obtain a stationary renewal process with $U^d(t) = t/\mu, t \geq 0$.

Here we consider another clever choice that can be used to illustrate the important renewal theorems. Let us choose $G(t) := F_0 * H(t)$, where $H(t)$ is a distribution function concentrated on $[0, \infty)$. In this case, we find that

$$\widehat{U}^d(s) = \frac{1}{\mu s}\widehat{H}(s),$$

and it follows that

$$U^d(t) = \frac{1}{\mu}\int_0^t H(x)dx.$$

From here it is easy to verify the important renewal theorems. For convenience, let $\overline{H}(t) = 1 - H(t)$ denote the tail of $H(t)$ and $\mu(H) = \int_0^\infty x\,dH(x) \leq \infty$.

- Since $H(t) \to 1, t \to \infty$, we have the elementary renewal theorem

$$\frac{1}{t}U^d(t) = \frac{1}{\mu t}\int_0^t H(u)du \to \frac{1}{\mu}, \quad t \to \infty.$$

Further on, we obtain Blackwell's theorem

$$U^d(t + h) - U^d(t) = \frac{1}{\mu}\int_0^h H(t + u)du \to \frac{h}{\mu}, \quad t \to \infty.$$

- If $\mu(H) < \infty$, we have

$$U^d(t) - \frac{t}{\mu} = \frac{1}{\mu}\int_0^t H(x)dx - \frac{1}{\mu}\int_0^t dx$$

$$= -\frac{1}{\mu}\int_0^t (1 - H(x))dx \to -\frac{1}{\mu}\int_0^\infty (1 - H(t))dt = -\frac{\mu(H)}{\mu}.$$

- If $\mu(H) < \infty$, we also obtain that

$$U^d(t) - \frac{t}{\mu} + \frac{\mu(H)}{\mu} = \frac{1}{\mu} \int_t^\infty \overline{H}(u) du.$$

Then if $\overline{H}(t) \in RV(-\beta)$, $\beta > 1$, we obtain that (see Theorem A.1)

$$U^d(t) - \frac{t}{\mu} + \frac{\mu(H)}{\mu} \sim \frac{t\overline{H}(t)}{\mu(\beta - 1)}, \quad t \to \infty.$$

- Let us turn to Blackwell's theorem. We have

$$U^d(t + h) - U^d(t) - \frac{h}{\mu} H(t) = \frac{1}{\mu} \int_0^h (H(t + u) - H(t)) du.$$

If $H(t)$ has a density $h(t)$ such that $h(\log t) \in RV(0)$, then we have

$$U^d(t + h) - U^d(t) - \frac{h}{\mu} H(t) = \frac{1}{\mu} \int_0^h \int_0^u h(t + y) dy du \sim \frac{h(t)}{\mu} \frac{h^2}{2}.$$

Rearranging we find also that

$$U^d(t + h) - U^d(t) - \frac{h}{\mu} + \frac{h}{\mu} \overline{H}(t) = \frac{1}{\mu} \int_0^h \int_0^u h(t + y) dy du \sim \frac{h(t)}{\mu} \frac{h^2}{2}.$$

- Suppose that the tail $\overline{H}(t)$ satisfies

$$\lim_{x \to \infty} \frac{\overline{H}(x + y)}{\overline{H}(x)} = \exp(\alpha y), \quad \text{for every } y.$$

Using

$$U^d(t + h) - U^d(t) - \frac{h}{\mu} = -\frac{1}{\mu} \int_0^h \overline{H}(t + u) du,$$

we obtain that

$$U^d(t + h) - U^d(t) - \frac{h}{\mu} \sim -\overline{H}(t) \frac{1}{\mu} \int_0^h \exp(\alpha u) du \sim -\overline{H}(t) \frac{1}{\mu} \frac{\exp(\alpha t) - 1}{\alpha}.$$

If $\alpha = 0$, we get that

$$U^d(t+h) - U^d(t) - \frac{h}{\mu} \sim -\overline{H}(t)\frac{h}{\mu}.$$

References

1. Alsmeyer, G.: Erneuerungstheorie. Teubner, Stuttgart (1991)
2. Asmussen, S.: Applied Probability and Queues, 2nd edn. Springer, New York (2003)
3. Choquet, G., Deny, J.: Sur l'equation de convolution $\mu = \mu * \sigma$. C. R. Acad. Sci. Paris **250**, 799–801 (1960)
4. Daley, D.J., Ver-Jones, D.: An Introduction to the Theory of Point Processes, vol. 1, 2nd edn. Springer, Berlin (2003)
5. Dippon, J., Walk, H.: Simplified analytical proof of Blackwell's renewal theorem. Stat. Probab. Lett. **74**, 15–20 (2005)
6. Feller, W.: An Introduction to Probability Theory and its Applications, vol. II. Wiley, New York (1971)
7. Kac, M.: A remark on Wienner's Tauberian theorem. Proc. Am. Math. Soc. **16**, 1155–1157 (1965)
8. Lindvall, T.: A probabilistic proof of Blackwell's renewal theorem. Ann. Probab. **5**(3), 482–485 (1977)
9. Ney, P.: A refinement of the coupling method in renewal theory. Stoch. Proc. Appl. **11**, 11–26 (1981)
10. Omey, E., Vesilo, R.: Local limit theorems for shock models. HUB Research paper 2011/23. HUB, Brussel (2011)
11. Serfozo, R.: Basic of Applied Stochastic Processes. Springer, New York (2009)
12. Smith, W.L.: Asymptotic renewal theorems. Proc. R. Soc. Edinb. Sect. A **64**, 9–48 (1954)

Chapter 2
Discrete Time Renewal Processes

Abstract In this chapter, we give a review of discrete renewal theory and prove the basic theorems for renewal sequences. We provide two different proofs of the theorem of Erdös-Feller-Pollard. Using extensions of a theorem of Wiener, we also obtain several rate of convergence results in discrete renewal theory.

Keywords Discrete time · Renewal sequence · Erdös-Feller-Pollard theorem · Lifetime processes · Rate of convergence

2.1 Introduction

In Chap. 1 we see that in the formulations of the main theorems one has to consider separately the cases of lattice and nonlattice distributions of the interarrival times. In the present chapter, we will consider discrete time renewal processes assuming that the interarrival times T_n have a lattice distribution with span $a = 1$. In other words, we assume that

$$T, \ T_1, \ T_2, \ \ldots, \ T_n, \ \ldots$$

are i.i.d. integer valued, nonnegative r.v.'s with distribution

$$\Pr\{T_n = t\} = p_t \geq 0, \ \ t = 0, 1, 2, \ldots,$$

such that $GCD\{t : p_t > 0\} = 1$, i.e. $\{p_t\}_{t=0}^{\infty}$ is nonperiodic.

Let us denote by

$$f(z) := \mathbf{E}[z^T] = \sum_{t=0}^{\infty} p_t z^t, \ \ |z| \leq 1$$

K. V. Mitov and E. Omey, *Renewal Processes*, SpringerBriefs in Statistics,
DOI: 10.1007/978-3-319-05855-9_2, © The Author(s) 2014

the probability generating function (p.g.f.) of the r.v. T. Evidently, under these assumptions the renewal sequence

$$S_0 = 0, \quad S_n = T_1 + T_2 + \ldots + T_n, \quad n = 1, 2, \ldots$$

takes values in nonnegative integers. Hence, the renewal counting processes

$$N(t) = \sup\{n \geq 0 : S_n \leq t\} \text{ and } M(t) = \inf\{n \geq 1 : S_n > t\}$$

can increase only at the integer time points $t = 0, 1, 2, \ldots$. So, they are discrete time stochastic processes. Denote by:

$$p_t(n) := \Pr\{S_n = t\} = p_t^{n*}, t = 0, 1, 2, \ldots, \quad n = 1, 2, \ldots$$

and

$$f_n(z) := \mathbf{E}[z^{S_n}] = \sum_{t=0}^{\infty} p_t(n) z^t, \quad |z| \leq 1, n = 1, 2, \ldots.$$

Let us recall that $p_t^{1*} = p_t$, $t = 0, 1, 2, \ldots$ and $p_t^{n*} = \sum_{k=0}^{t} p_{t-k}^{(n-1)*} p_k^{1*}$ for $t = 0, 1, 2 \ldots$ and $n = 2, 3, \ldots$.

Clearly, $f_n(z) = f(z)^n, n = 1, 2, 3, \ldots$.

Let us recall that the path properties of the process $N(t)$ do not depend on the fact that the distribution of interarrival times is lattice or not. So we will restrict ourselves only to represent the discrete time versions of the main limit theorems.

2.2 Theorem of Erdös, Feller and Pollard

Similarly to the continuous time case, we are interested in the behavior or the renewal function $U(t)$ as $t \to \infty$. Let us note that now the renewal events can occur only in the integer time points $t = 0, 1, 2, \ldots$. Therefore, it makes sense to study the measure $U(\{t\})$, that is the mean number of the renewals at time t. In other words, if we define the random variable $I_t = 1$, if there is a renewal event at time t and $I_t = 0$ otherwise then $U(\{t\}) = \mathbf{E}[I_t]$. Let us denote

$$u_t = \Pr\{I_t = 1\}, \quad 1 - u_t = \Pr\{I_t = 0\}$$

for $t = 0, 1, 2, \ldots$ then

$$U(\{t\}) = \mathbf{E}[I_t] = u_t.1 + (1 - u_t).0 = u_t, \ t = 0, 1, 2, \ldots. \tag{2.1}$$

Since we count $t = 0$ as a renewal then we will assume that $u_0 = 1$. From the definition of u_t it follows that

$$u_t = \sum_{n=0}^{\infty} \Pr\{S_n = t\} = \sum_{n=0}^{\infty} p_t^{n*}, \tag{2.2}$$

and

$$U(t) = U([0, t]) = \mathbf{E}\left[\sum_{k=0}^{t} I_k\right] = \sum_{k=0}^{t} u_k = \sum_{k=0}^{t}\sum_{n=0}^{\infty} \Pr\{S_n = k\}$$

$$= \sum_{n=0}^{\infty}\sum_{k=0}^{t} \Pr\{S_n = k\} = \sum_{n=0}^{\infty} \Pr\{S_n \le t\} = \mathbf{E}[N(t)] + 1 = \mathbf{E}[M(t)].$$

Multiplying both sides of the Eq. (2.2) by z^t, ($|z| \le 1$) and summing on $t = 0, 1, 2, \ldots$, we get

$$u(z) = \sum_{t=0}^{\infty}\sum_{k=0}^{\infty} \Pr\{S_k = t\}z^t = \sum_{k=0}^{\infty}\sum_{t=0}^{\infty} \Pr\{S_k = t\}z^t$$

$$= \sum_{k=0}^{\infty} f^k(z) = \frac{1}{1 - f(z)}. \tag{2.3}$$

From here it follows that $u(z) = 1 + u(z)f(z)$ and

$$u_0 = 1, \quad u_t = p_t u_0 + p_{t-1} u_1 + \ldots + p_1 u_{t-1} \text{ for } t = 1, 2, \ldots,$$

which is the discrete time version of the renewal equation (1.6).

Example 2.1 Let T be given by $\Pr\{T = 1\} = p$ and $\Pr\{T = 2\} = q = 1 - p$, where $0 < p < 1$. Then $\mu = \mathbf{E}[T] = p + 2q = 1 + q$. In this case we obtain

$$u_0 = 1, \quad u_1 = p, \quad u_t = pu_{t-1} + qu_{t-2}, t \ge 2.$$

The generating function of u_t is given by

$$u(z) = \frac{1}{1 - pz - qz^2} = \frac{1}{p(1 - z) + q(1 - z^2)} = \frac{1}{(1 - z)(1 + qz)}$$

$$= \frac{1}{1 + q}\frac{1}{1 - z} + \frac{q}{1 + q}\frac{1}{1 + qz},$$

for $|z| \le 1$. From here we find that

$$u_t = \frac{1}{1 + q} + \frac{q}{1 + q}(-q)^t, \quad t \ge 1.$$

Note that

$$u_t \to \frac{1}{1+q} = \frac{1}{\mu}, \quad t \to \infty.$$

The same limit holds true in the general case. The result was proved by Erdös, Feller, and Pollard at 1949 (see [3]) and then it was generalized by Blackwell for the continuous time.

Theorem 2.1 (Erdös, Feller, and Pollard) *Assume that $\mu = \mathbf{E}[T] \leq \infty$. Then*

$$\lim_{t \to \infty} u_t = \frac{1}{\mu},$$

where $1/\mu$ is defined to be zero if $\mu = \infty$.

Remark 2.1 Here, we give two proofs only in the case when $\mu < \infty$. The first one is based on a result of Wiener (see e.g. Rudin [7], Theorem 18.21) and is close to the first proof in the original paper of Erdös, Feller and Pollard. The second proof is based on the coupling method.

Lemma 2.1 (Rudin [7], Theorem 18.21) *Suppose that the sequence $\{x_n\}_{n=0}^{\infty}$ is such that $x(z) = \sum_{n=0}^{\infty} x_n z^n$ is absolutely convergent for $|z| \leq 1$ and that $\sum_{n=0}^{\infty} |x_n| < \infty$. If $x(z) \neq 0$ for every $z, |z| \leq 1$, then*

$$\frac{1}{x(z)} = \sum_{n=0}^{\infty} y_n z^n$$

is absolutely convergent for $|z| \leq 1$ and

$$\sum_{n=0}^{\infty} |y_n| < \infty.$$

Proof I of Theorem 2.1. Define the sequence $r_t = \Pr\{T > t\} \, t = 0, 1, 2, \ldots$ and its generating function $r(z) = \sum_{t=0}^{\infty} r_t z^t$, $|z| \leq 1$. The following representation holds true

$$r(z) = \frac{1 - f(z)}{1 - z}.$$

Now note that $r(z) \neq 0$ for $|z| \leq 1$ and $z \neq 1$. For $z = 1$, we have $r(1) = f'(1) = \mu \neq 0$. From the result of Wiener we get that

$$\Lambda(z) = \frac{1}{r(z)} = \sum_{t=0}^{\infty} \Lambda_t z^t \quad \text{and} \quad \sum_{t=0}^{\infty} |\Lambda_t| < \infty.$$

It follows that $(1 - z)u(z) = \Lambda(z)$ and hence also that

$$u_0 = \Lambda_0, \quad \text{and} \quad (u_t - u_{t-1}) = \Lambda_t, \quad t \geq 1.$$

Taking sums, we get that

$$u_0 + \sum_{k=1}^{t}(u_k - u_{k-1}) = \sum_{k=0}^{t} \Lambda_k$$

or equivalently $u_t = \sum_{k=0}^{t} \Lambda_k$. Since $\sum_{k=0}^{\infty} \Lambda_k = \Lambda(1) = 1/r(1) = 1/\mu$, we get that

$$u_t \to 1/\mu, \quad \text{as} \quad t \to \infty,$$

which completes the proof of the theorem.   △

Proof II of Theorem 2.1.(Barbu and Limnios [1], Chap. 2). As before, let $\{S_n, n \geq 0\}$ denote the renewal sequence with $S_0 = 0$ and $S_n = \sum_{i=1}^{n} T_i$, $n \geq 1$. We assume that $\Pr\{T < \infty\} = 1$ and that $\mu = \mathbf{E}[T] < \infty$. Let $\{S_n^*, \ n \geq 0\}$ denote the stationary renewal sequence associated with $\{S_n, n \geq 0\}$. More precisely, we have

$$S_n^* = T_0^* + \sum_{i=1}^{n} T_i^*, \ n \geq 0,$$

where the T_i^*, $i \geq 1$ are i.i.d. with $T_i^* \overset{d}{=} T$ and independent of the T_i^*,

$$\Pr\left\{T_0^* = n\right\} = \Pr\{T > n\}/\mu.$$

Let u_t and u_t^* denote the corresponding renewal sequences. Because of the definition, we have

$$u_t^* = \Pr\left\{S_k^* = t \text{ for some } k \in \mathbf{N}\right\} = 1/\mu.$$

Now, define $U_0 = T_0 - T_0^*, U_n = U_{n-1} + (T_n - T_n^*) = S_n - S_n^*, n \geq 0$ and denote by N the first time that we have the same number of renewals in the two processes, i.e.

$$N = \min\{n : U_n = 0\}.$$

Since $\mathbf{E}[T_i] = \mathbf{E}[T_i^*] = \mu$, then $\mathbf{E}[T_i - T_i^*] = 0$. Hence $U_n, n = 0, 1, 2 \ldots$ is a recurrent Markov chain. Consequently, $\Pr\{N < \infty\} = 1$.

Thus, for $n \geq N$ we have that $S_n \overset{d}{=} S_n^*$. Now we find

$$u_t = \Pr\{S_k = t \text{ for some } k \in \mathbf{N}\}$$
$$= \Pr\{S_k = t \text{ for some } k \geq N\} + \Pr\{S_k = t \text{ for some } k < N\}$$
$$= \Pr\left\{S_k^* = t \text{ for some } k \geq N\right\} + \Pr\{S_k = t \text{ for some } k < N\}$$
$$= 1/\mu - \Pr\left\{S_k^* = t \text{ for some } k < N\right\} + \Pr\{S_k = t \text{ for some } k < N\}.$$

Now note that $\{S_k = t \text{ for some } k < N\} \subseteq \{S_N \geq t\}$. We have

$$\lim_{t \to \infty} \Pr\{S_k = t \text{ for some } k < N\} \leq \Pr\{S_N = \infty\} = 0.$$

In a similar way, we have

$$\lim_{t \to \infty} \Pr\{S_k^* = t \text{ for some } k < N\} \leq \Pr\{S_N^* = \infty\} = 0.$$

It follows that $u_t \to 1/\mu$. $\triangle$

Remark 2.2 The general proof of the theorem given in the paper Erdös et al.[3] does not depend on the fact that $\mu < \infty$. It can be found also in Feller [4].

2.3 Rate of Convergence in Discrete Renewal Theorem

Following Rogozin [5], we define three classes of sequences.

$$D = \left\{ \{x_n\} : x(z) = \sum_{n=0}^{\infty} x_n z^n \text{ is absolutely convergent for } |z| \leq 1 \right\};$$

$$R_1(\alpha, L) = \left\{ \{x_n\} \in D : x_n = O(n^{-\alpha} L(n)) \right\};$$

$$R_2(\alpha, L) = \left\{ \{x_n\} \in D : x_n = o(n^{-\alpha} L(n)) \right\},$$

where $L(.)$ denotes a slowly varying function at infinity.

Rogozin proves the following Wiener type of results.

Lemma 2.2 (Rogozin [6], p. 664) (i) *Suppose that* $\{x_n\} \in R_1(\alpha, L)$ *and* $x(z) \neq 0$ *for* $|z| \leq 1$. *Then*

$$\frac{1}{x(z)} = y(z) = \sum_{n=0}^{\infty} y_n z^n$$

and $\{y_n\} \in R_1(\alpha, L)$.

(ii) *Suppose that* $\{x_n\} \in R_2(\alpha, L)$, $\alpha > 1$ *and* $x(z) \neq 0$ *for* $|z| \leq 1$. *Then*

$$\frac{1}{x(z)} = y(z) = \sum_{n=0}^{\infty} y_n z^n$$

and $\{y_n\} \in R_2(\alpha, L)$.

By replacing $\Phi(x) = 1/x$ by other suitable functions, we get the following general result.

Lemma 2.3 (Rogozin [5], Theorem 5) (i) *Suppose that $\{x_n\} \in R_1(\alpha, L)$ and suppose that $\Phi(z)$ is a complex function that is analytic in a region that contains the set $\{x(z) : |z| \le 1\}$. Then*

$$q(z) = \Phi(x(z)) = \sum_{n=0}^{\infty} q_n z^n$$

and $\{q_n\} \in R_1(\alpha, L)$.
  (ii) *A similar result holds for $R_2(\alpha, L)$ with $\alpha > 1$.*

The next lemma states the conditions under which one obtains an asymptotic equality.

Lemma 2.4 (Borovkov [2], p. 258, Theorem 1) *Suppose that $\{x_n\} \in R_1(\alpha, L)$ and that*

$$0 < An^{-\alpha}L(n) \le |x_n| \le Bn^{-\alpha}L(n), \quad n \ge 1.$$

With Φ as in the previous lemma, we have

$$q(z) = \Phi(x(z)) = \sum_{n=0}^{\infty} q_n z^n$$

and $q_n \sim x_n \Phi'(x(1))$ as $n \to \infty$.

Now we are ready to formulate the results for the rate of convergence in the renewal theorem. The results depend on the behavior of the tail sequence $r_n = \Pr\{T > n\}$. Using Lemma 2.3 and 2.4, we obtain the following result.

Theorem 2.2 *Suppose that $\alpha > 1$, $L(x) \in RV(0)$ and $\mu < \infty$. Then*

  (i) $r_n = O(n^{-\alpha}L(n)) \Longleftrightarrow u_n - u_{n-1} = O(n^{-\alpha}L(n))$;
  (ii) $r_n = o(n^{-\alpha}L(n)) \Longleftrightarrow u_n - u_{n-1} = o(n^{-\alpha}L(n))$;
  (iii) $r_n \sim n^{-\alpha}L(n) \Longleftrightarrow \mu^2(u_n - u_{n-1}) \sim -n^{-\alpha}L(n)$.

Proof As in the first proof of Theorem 2.1, we use the equation

$$r(z) = \frac{1 - f(z)}{1 - z}, \text{ and } r(1) = \mu.$$

Then by Lemma 2.1 we get

$$\Lambda(z) = \frac{1}{r(z)} = \sum_{n=0}^{\infty} \Lambda_n z^n$$

where $\sum_0^\infty |\Lambda_n| < \infty$. Since $\Lambda(z) = (1-z)u(z)$, then $\Lambda_0 = u_0$ and $\Lambda_n = u_n - u_{n-1}, n \geq 1$. Now (i) and (ii) follow from these relations and Lemma 2.3.

(iii) First suppose that $r_n \sim n^{-\alpha} L(n)$. Using Lemma 2.4 with $\Phi(z) = 1/z$, it follows that

$$\Lambda_n \sim r_n \Phi'(r(1)) = -r_n/\mu^2.$$

Since $\Lambda_n = (u_n - u_{n-1})$, the result follows. The converse follows in a similar way, which completes the proof of the theorem.   △

From Theorem 2.2 (iii), we have

$$\mu^2(u_n - u_{n-1}) \sim -r_n.$$

By taking sums, we have

$$-(1+\varepsilon)\sum_{n=k}^{K} r_n \leq \mu^2 \sum_{n=k}^{K}(u_n - u_{n-1}) \leq -(1-\varepsilon)\sum_{n=k}^{K} r_n$$

or equivalently

$$-(1+\varepsilon)\sum_{n=k}^{K} r_n \leq \mu^2(u_K - u_{k-1}) \leq -(1-\varepsilon)\sum_{n=k}^{K} r_n.$$

Taking limits with respect to K, we obtain that

$$-(1+\varepsilon)\sum_{n=k}^{\infty} r_n \leq \mu^2(1/\mu - u_{k-1}) \leq -(1-\varepsilon)\sum_{n=k}^{\infty} r_n.$$

Using $r_n \sim n^{-\alpha} L(n), \alpha > 1$, it follows that

$$\sum_{n=k}^{\infty} r_n \sim \frac{1}{\alpha - 1} k r_k$$

The final conclusion is that

$$u_n - \frac{1}{\mu} \sim \frac{1}{\mu^2(\alpha - 1)} n r_n.$$

Under the conditions of Theorem 2.2 (i), (ii) in a similar way we obtain $O(.)$ and $o(.)$ results.

Corollary 2.1 *Suppose that $\alpha > 1$, $L(x) \in RV(0)$ and $\mu < \infty$. Then*

(i) $r_n = O(n^{-\alpha}L(n)) \Longrightarrow u_n - 1/\mu = O(n^{1-\alpha}L(n))$;
(ii) $r_n = o(n^{-\alpha}L(n)) \Longrightarrow u_n - 1/\mu = o(n^{1-\alpha}L(n))$;
(iii) $r_n \sim n^{-\alpha}L(n) \Longrightarrow \mu(u_n - 1/\mu) \sim n^{1-\alpha}L(n)/(\alpha - 1)$.

2.4 Lifetime Processes

Let us have a look at lifetime processes $A(t) = t - S_{N(t)}$, $B(t) = S_{M(t)} - t$ and $C(t) = T_{M(t)}$. Now they are discrete time integer valued processes. We will derive the representation for their probability distributions. Then we will prove limit theorems.

For the distribution of $A(t)$, we get for $n \geq 0$, $t \geq n$,

$$\Pr\{A(t) = n\} = \Pr\{S_{N(t)} = t - n\}$$

$$= \sum_{k=0}^{\infty} \Pr\{S_{N(t)} = t - n, N(t) = k\}$$

$$= \sum_{k=0}^{\infty} \Pr\{S_k = t - n, S_k \leq t < S_{k+1}\}$$

$$= \sum_{k=0}^{\infty} \Pr\{S_k = t - n, t - n \leq t < t - n + T_{k+1}\}$$

$$= \sum_{k=0}^{\infty} \Pr\{S_k = t - n\}\Pr\{n < T_{k+1}\}$$

$$= \Pr\{T > n\}\sum_{k=0}^{\infty} \Pr\{S_k = t - n\}$$

$$= u_{t-n}\Pr\{T > n\}.$$

Consider now the residual lifetime $B(t) = S_{M(t)} - t$. By conditioning on $M(t)$, for $n \geq 1$, $t \geq 0$, we have

$$\Pr\{B(t) = n\} = \sum_{i=1}^{\infty} \Pr\{S_i = t + n, M(t) = i\} = \sum_{i=1}^{\infty} \Pr\{S_i = t + n, S_{i-1} \leq t < S_i\}$$

$$= \Pr\{T = t + n\} + \sum_{i=2}^{\infty}\sum_{j=1}^{t} \Pr\{T_i = t + n - j\}\Pr\{S_{i-1} = j\}$$

$$= \Pr\{T = t + n\} + \sum_{j=1}^{t} \Pr\{T = t + n - j\}\sum_{i=2}^{\infty} \Pr\{S_{i-1} = j\}$$

$$= \Pr\{T = t + n\} + \sum_{j=1}^{t} \Pr\{T = t + n - j\} u_j$$

$$= \sum_{j=0}^{t} \Pr\{T = t + n - j\} u_j.$$

For the distribution of $C(t)$, we get

$$\Pr\{C(t) = n\} = \Pr\{T_{N(t)+1} = n\}$$

$$= \sum_{j=0}^{\infty} \Pr\{T_{j+1} = n, N(t) = j\} = \sum_{j=0}^{\infty} \Pr\{T_{j+1} = n, S_j \le t < S_j + T_{j+1}\}$$

$$= \sum_{j=0}^{\infty} \sum_{k=0}^{\infty} \Pr\{T_{j+1} = n, S_j \le t < S_j + T_{j+1}, S_j = k\}$$

$$= \sum_{j=0}^{\infty} \sum_{k=0}^{\infty} \Pr\{T_{j+1} = n, k \le t < k + n, S_j = k\}$$

$$= \sum_{j=0}^{\infty} \sum_{k=t-n+1}^{t} \Pr\{T = n\} \Pr\{S_j = k\}$$

$$= \Pr\{T = n\} \sum_{k=t-n+1}^{t} u_k.$$

Now we are ready to prove the following limit theorem.

Theorem 2.3 *Assume that $\mu < \infty$. Then*

$$\lim_{t \to \infty} \Pr\{A(t) = n\} = \frac{1}{\mu} \Pr\{T > n\}.$$

$$\lim_{t \to \infty} \Pr\{B(t) = n\} = \frac{1}{\mu} \Pr\{T \ge n\}.$$

$$\lim_{t \to \infty} \Pr\{C(t) = n\} = \frac{n}{\mu} \Pr\{T = n\}.$$

Proof Using renewal theorem, we easily obtain the first and the third limits from $\Pr\{A(t) = n\} = u_{t-n} \Pr\{T > n\}$ and $\Pr\{C(t) = n\} = \Pr\{T = n\} \sum_{k=t-n+1}^{t} u_k$, respectively

The proof of the second relation is not so evident. For $t > t°$ and $t°$ sufficiently large, we have

$$\Pr\{B(t) = n\} = \left(\sum_{j=0}^{t^\circ} + \sum_{j=t^\circ+1}^{t} \right) \Pr\{T = t + n - j\} u_j$$

$$= J_1(t) + J_2(t).$$

In $J_2(t)$, we have

$$\left(\frac{1}{\mu} - \varepsilon \right) \Pr\{n \le T \le t + n - t^\circ - 1\} \le J_2(t) \le \left(\frac{1}{\mu} + \varepsilon \right) \Pr\{n \le T \le t + n - t^\circ - 1\},$$

and it follows that

$$\left(\frac{1}{\mu} + \varepsilon \right) \Pr\{n \le T\} \le \liminf_{n \to \infty} J_2(t) \le \limsup_{n \to \infty} J_2(t) \le \left(\frac{1}{\mu} + \varepsilon \right) \Pr\{n \le T\}.$$

For $J_1(t)$, we have

$$J_1(t) \le \left(\sup_{j \le t^\circ} u_j \right) \Pr\{t + n - k^\circ \le T \le t + n\} \to 0, \text{ as } t \to \infty.$$

This completes the proof of the theorem.   △

Remark 2.3 This theorem is the discrete time analog of Theorem 1.18.

Let us note that we derive the equations for the distributions of the lifetime processes conditioning on the last renewal event before time t. As in the continuous time case these distributions are solutions of three renewal equations. We will derive one of them using the standard renewal argument.

Consider the residual lifetime $B(t) = S_{M(t)} - t$. Note that $B(0) = T_1$. By conditioning on T_1, we have

$$\Pr\{B(t) > n\} = \sum_{i=1}^{\infty} \Pr\{B(t) > n \mid T_1 = i\} p_i.$$

Now note that the process restarts at T_1 and we get

$$\Pr\{B(t) > n \mid T_1 = i\} = 1, \quad \text{if } i > t + n$$
$$\Pr\{B(t) > n \mid T_1 = i\} = 0, \quad \text{if } t < i \le t + n$$
$$\Pr\{B(t) > n \mid T_1 = i\} = \Pr\{B(t - i) > n\} \quad \text{if } i \le t.$$

It follows that

$$\Pr\{B(t) > n\} = \Pr\{T > t + n\} + \sum_{i=0}^{t} \Pr\{B(t - i) > n\} p_i, t \ge 0. \qquad (2.4)$$

This renewal equation can be solved by using generating functions. Let

$$p_n(z) = \sum_{t=0}^{\infty} \Pr\{B(t) > n\} z^t$$

$$g(z) = \sum_{t=0}^{\infty} \Pr\{T > t + n\} z^t.$$

Multiplying Eq. (2.4) by z^t and summing on $t = 0, 1, 2, \ldots, \infty$, we get that

$$p_n(z) = g(z) + f(z) p_n(z).$$

Using also (2.3) we get

$$p_n(z) = \frac{g(z)}{1 - f(z)} = g(z) u(z).$$

Therefore we get the following representation for the tail of the distribution of $B(t)$,

$$\Pr\{B(t) > n\} = \sum_{i=0}^{t} \Pr\{T > t + n - i\} u_i.$$

In a similar way one can derive the renewal equations for the distributions of the processes $A(t)$ and $C(t)$.

We conclude this chapter with two examples.

Example 2.2 Let T be a Bernoulli random variable with $\Pr\{T = 1\} = p$ and $\Pr\{T = 0\} = q = 1 - p$. Then $f(z) = \mathbf{E}[z^T] = f(z) = q + pz$ and $\mu = p$. The renewal measure has generating function

$$u(z) = \frac{1}{1 - f(z)} = \frac{1}{p(1 - z)}$$

and it follows that $u_t = 1/p = 1/\mu$ for every $t = 1, 2, \ldots$.

Example 2.3 Let T have a geometric distribution $\Pr\{T = k\} = pq^{k-1}, k \geq 1$. We have $f(z) = \mathbf{E}[z^T] = pz/(1 - qz)$ and $\mu = \mathbf{E}[T] = 1/p$. The renewal measure has generating function

$$u(z) = \frac{1}{1 - f(z)} = \frac{1 - qz}{1 - qz - pz} = \frac{1 - qz}{1 - z},$$

which can be rewritten as

$$u(z) = p \frac{1}{1-z} + q.$$

Hence, $u_0 = 1$ and $u_t = p = 1/\mu$ for every $t = 1, 2, \dots$.

References

1. Barbu, V.S., Limnios, N.: Semi-Markov Chains and Hidden Semi-Markov Models Toward Applications. Springer, New York (2008)
2. Borovkov, A.A.: Stochastic Processes in Queueing Theory. Springer, New York (1976)
3. Erdös, P., Feller, W., Pollard, H.: A property of power series with positive coefficients. Bull. Am. Math. Soc. **55**(2), 201–204 (1949)
4. Feller, W.: An Introduction to Probability Theory and its Applications, vol. I. Wiley, New York (1968)
5. Rogozin, B.A.: Asymptotics of the coefficients in the Lévy-Wiener theorems on absolutely convergent trigonometric series. Sib. Math. J. **14**, 1304–1312 (1973)
6. Rogozin, B.A.: An estimate of the remainder term in limit theorems of renewal theory. Theory Prob. Appl. **18**, 662–677 (1973)
7. Rudin, W.: Real and Complex Analysis. McGraw-Hill, New York (1970)

Chapter 3
Extensions and Applications

Abstract This chapter is devoted to some extensions and applications of renewal theory. First, we discuss the renewal theorems in the case where the underlying mean is infinite. We proceed by a short discussion of alternating renewal processes. In order to discuss renewal reward processes and superposed renewal processes, we need some basic properties of bivariate renewal theory. The chapter ends with some important applications of renewal theory in different areas of stochastic processes.

Keywords Infinite mean interarrival times · Renewal theorems · Lifetime processes · Dynkin-Lamperti theorem · Superposition of renewal processes · Alternating renewal processes · Renewal reward processes · Queuing models · Insurance models

3.1 Renewal Processes with Infinite Mean of Interarrival Times

The theorems stated in the previous sections do not give precise information about the asymptotic of the renewal functions and the limiting distributions of the lifetime processes in case when $\mu = \infty$. It was mentioned only that in case $\mu = \infty$, the limit $1/\mu$ has to be considered as 0. So in the elementary renewal theorem if $\mu < \infty$, then

$$U(t) \sim t/\mu, \quad t \to \infty,$$

i.e., the mean number of renewal events increases linearly with time. On the other hand, if $\mu = \infty$ we have only that

$$U(t)/t \to 0, \quad t \to \infty.$$

This relation means that in case of infinite mean the rate of change of the renewal function is slower than linear, $U(t) = o(t), t \to \infty$.

K. V. Mitov and E. Omey, *Renewal Processes*, SpringerBriefs in Statistics,
DOI: 10.1007/978-3-319-05855-9_3, © The Author(s) 2014

The situation in the other limit theorems proved in Chap. 1 is similar. Let us recall that the limit distributions of the lifetimes are degenerate. If we interpret $1/\mu = 0$, when $\mu = \infty$ then

$$\lim_{t \to \infty} \Pr\{A(t) \le x\} = \frac{1}{\mu} \int_0^x (1 - F(u))du = 0.$$

This means that without normalization $A(t) \to \infty$ in distribution. In this section, we represent the main results of renewal theory in the infinite mean case following the structure of Chap. 1.

3.1.1 Basic Conditions

Throughout this section we suppose that $\{S_n, \ n = 0, 1, 2 \ldots\}$ is a pure renewal process and the distribution function $F(t)$ of interarrival times satisfies the following conditions:

(a) $F(t)$ is nonlattice and $F(0) = 0$,
(b) $F(t)$ has a regularly varying tail with exponent $-\beta$, $\beta \in [0, 1]$, that is

$$1 - F(t) \sim \frac{t^{-\beta} L(t)}{\Gamma(1 - \beta)}, \quad t \to \infty, \tag{3.1}$$

where $L(\cdot)$ is a slowly varying function at infinity.

The factor $[\Gamma(1 - \beta)]^{-1}$ is taken for convenience.

Obviously, if $\beta < 1$ then $\mu = \infty$. In the case where $\beta = 1$, it is possible to have either $\mu = \infty$ or $\mu < \infty$ depending on the s.v.f. $L(t)$. Since the case $\mu < \infty$ has been considered in the previous sections we assume that

$$\mu(t) := \int_0^t (1 - F(u))du \to \infty \quad \text{as} \quad t \to \infty. \tag{3.2}$$

Under these assumptions the following theorem holds.

Theorem 3.1 (i) *Assume* $\beta < 1$ *Then* (3.1) *is equivalent to*

$$\mu(t) \sim \frac{t^{1-\beta} L(t)}{\Gamma(2 - \beta)}, \quad t \to \infty.$$

(ii) *If* (3.1) *holds with* $\beta = 1$ *then* $\mu(t) \uparrow \infty$ *and it varies slowly at infinity.*

The proof follows from Theorems A.1 and A.2.

Remark 3.1 We will denote by

$$\Gamma(p) = \int_0^\infty y^{p-1} e^{-x} dx, \quad Rep > 0,$$

$$B(p, q) = \int_0^1 y^{p-1}(1 - y)^{q-1} dx, \quad Rep > 0, \quad Req > 0,$$

and

$$B_x(p, q) = \int_0^x y^{p-1}(1 - y)^{q-1} dy, \quad Rep > 0, Req > 0, \quad x \in (0, 1)$$

the Euler's gamma, beta, and incomplete beta function, respectively.

3.1.2 Limit Theorems for the One-Dimensional Distributions of $N(t)$

The following theorem is an analog of Theorem 1.3.

Theorem 3.2 *Assume that the distribution function of the interarrival times satisfies* (3.1) *with* $\beta \in (0, 1)$. *Then*

$$\lim_{t \to \infty} \Pr \left\{ (1 - F(t)) N(t) \geq \frac{x^{-\beta}}{\Gamma(1 - \beta)} \right\} \to G_\beta(x),$$

where $G_\beta(.)$ *is the stable distribution determined in Theorem A.5.*

Proof Let $x > 0$ be fixed. Under the conditions of the theorem we have (cf. Theorems A.5 and A.6), $\Pr\{S_n/a_n \leq x\} \to G_\beta(x)$ where $\{a_n\} \in RV(1/\beta)$ is determined by $n(1 - F(a_n)) \to \frac{1}{\Gamma(1-\beta)}$. Clearly, we have

$$\Pr\{N(a_n) \geq ny\} = \Pr \left\{ S_{[ny]} \leq a_{[ny]} \frac{a_n}{a_{[ny]}} \right\} \to G_\beta(y^{-1/\beta}),$$

locally uniformly in $y > 0$. Now choose t and n such that $a_n \leq t < a_{n+1}$. We have $N(a_n) \leq N(t) \leq N(a_{n+1})$. Moreover, we have $1 - F(a_n) \geq 1 - F(t) \geq 1 - F(a_{n+1})$ and consequently also that

$$n(1 - F(t)) \to \frac{1}{\Gamma(1 - \beta)}, \quad \text{as} \quad n, t \to \infty.$$

Using

$$\Pr\left\{ N(a_{n+1}) \geq (n+1)y\frac{n}{n+1} \right\} \leq \Pr\{N(t) \geq ny\} \leq \Pr\{N(a_n) \geq ny\}$$

it follows that

$$\Pr\{N(t) \geq ny\} \to G_\beta(y^{-1/\beta}),$$

locally uniformly in $y > 0$. Using $n(1 - F(t)) \to \frac{1}{\Gamma(1-\beta)}$, the result follows. $\triangle$

3.1.3 Elementary Renewal Theorem. Blackwell's Theorem. Key Renewal Theorems

3.1.3.1 Elementary Renewal Theorem

If $\beta \in [0, 1)$ then from (3.1) we get in terms of Laplace transforms the following equivalent relation (see Theorem A.4)

$$1 - \widehat{F}(s) \sim s^\beta L(1/s), \quad s \downarrow 0, \tag{3.3}$$

For the renewal function $\widehat{U}(s) = 1/(1 - \widehat{F}(s))$ Theorem A.4 shows that this is equivalent to

$$U(t) \sim \frac{t^\beta}{L(t)\Gamma(1 + \beta)}, \quad t \to \infty. \tag{3.4}$$

Therefore

$$\{1 - F(t)\} U(t) \to \frac{1}{\Gamma(1 - \beta)\Gamma(1 + \beta)} = \frac{\sin \pi\beta}{\pi\beta}, \quad t \to \infty.$$

In this way, we proved the analog of the elementary renewal theorem.

Theorem 3.3 *Suppose that (a), (b), and (3.2) hold. Then each of the following statements (i) and (ii) implies the other and both imply (3.5).*
   (i) $\mu(t)$ is regularly varying with exponent $1-\beta$.
   (ii) $U(t)$ is regularly varying with exponent β.

$$U(t) \sim C_\beta \frac{t}{\mu(t)}, \quad \text{as} \quad t \to \infty, \tag{3.5}$$

where $C_\beta = [\Gamma(\beta + 1)\Gamma(2 - \beta)]^{-1}$.

3.1.3.2 An Analog of Blackwell's Renewal Theorem

If $\mu = \infty$ then Blackwell's renewal theorem stated that for every fixed $h > 0$

$$U(t) - U(t - h) \to 0, \quad t \to \infty.$$

The following theorem has been proved by Erickson [13]. By our knowledge, it is the strongest result in the infinite mean case published by now.

Theorem 3.4 (Erickson [13]) *Suppose that (a), (b), and (3.2) hold true. If $\beta \in (\frac{1}{2}, 1]$, then for every fixed $h > 0$*

$$\lim_{t \to \infty} \mu(t)[U(t) - U(t - h)] = C_\beta h. \tag{3.6}$$

If $\beta \in (0, \frac{1}{2}]$, then for every fixed $h > 0$

$$\liminf_{t \to \infty} \mu(t)[U(t) - U(t - h)] = C_\beta h.$$

where $C_\beta = [\Gamma(\beta + 1)\Gamma(2 - \beta)]^{-1}$.

The next example shows that in special cases (3.6) holds true for every $0 < \beta < 1$.

Example 3.1 The Mittag–Leffler function

$$E_\beta(x) = \sum_{k=0}^{\infty} \frac{1}{\Gamma(1 + \beta k)} x^k$$

has been investigated by many authors (see, e.g., Feller [14], or more recently Pillai [29], Jayakumar and Suresh [22]). It is well known that $F(t) = 1 - E_\beta(-t^\beta)$, for $0 < \beta < 1$, is a distribution function with Laplace transform

$$\widehat{F}(s) = \frac{1}{1 + s^\beta}.$$

For the corresponding renewal function, we find that

$$\widehat{U}(s) = \frac{1}{1 - \widehat{F}(s)} = s^\beta.$$

Therefore,

$$U(t) = \frac{1}{\Gamma(1 + \beta)} t^\beta,$$

which gives that as $t \to \infty$,

$$U(t) - U(t-h) = \frac{1}{\Gamma(1+\beta)}(t^\beta - (t-h)^\beta) \sim \frac{1}{\Gamma(1+\beta)}\beta t^{\beta-1}h,$$

for every $\beta \in (0,1)$.

3.1.3.3 Key Renewal Theorems

Now we turn to the analogs of Smith's key renewal theorem (Theorem 1.12). They determine the asymptotic behavior of the solution of the renewal Eq. (1.31) in the infinite mean case ($\mu = \infty$). The first one is proved by Erickson in [13]. The second is proved by Anderson and Athreya [2].

Theorem 3.5 *Assume that $z(t)$ is dRi and $z(t) = O(t^{-1})$, $t > 0$. If $F(t)$ satisfies the condition (a), (b) and (3.2) with $\beta \in (\frac{1}{2}, 1]$, then*

$$Z(t) = U * z(t) \sim \frac{C_\beta}{\mu(t)} \int_0^\infty z(u)du, \quad t \to \infty.$$

The proof can be found in Erickson [13].

Theorem 3.6 *Assume that $F(t)$ satisfies (a), (b) and (3.2) with $\beta \in (\frac{1}{2}, 1]$. If $z(t):\mathbf{R}^+ \to \mathbf{R}^+$ is nonincreasing and such that $z(0) < \infty$, and*

$$z(t) = t^{-\gamma}L_1(t), \quad 0 \le \gamma < 1,$$

where $L_1(t)$ is a function slowly varying at infinity. Then

$$Z(t) = U * z(t) \sim C(\beta, \gamma) \frac{\int_0^t z(u)du,}{\mu(t)} \quad t \to \infty,$$

where $C(\beta, \gamma) = [(2-\gamma)B(\beta - \gamma + 1, 2 - \beta)]^{-1}$.

Proof (Anderson and Athreya [2]).
   1. Under the conditions of the theorem we have:

$$U(t) \sim C_\beta \frac{t}{\mu(t)}, \quad t \to \infty, \tag{3.7}$$

$\mu(t)$ varies regularly with exponent $1 - \beta$,

$$\lim_{t\to\infty} \mu(t)(U(t+h) - U(t)) = C_\beta h,$$

$$\int_0^t z(u)du \sim \frac{tz(t)}{1-\gamma}, \tag{3.8}$$

where $C_\beta = [\Gamma(1+\beta)\Gamma(2-\beta)]^{-1}$.

2. Let $\varepsilon > 0$ be fixed. Choose $0 < \delta < 1$ such that

$$\beta B(\beta, 1 - \gamma) - \beta \int_0^\delta (1 - u)^{-\gamma} u^{\beta - 1} du < \varepsilon \tag{3.9}$$

Subdivide the integral into three parts

$$U * z(t) = \int_0^t z(t - u) dU(u) = \int_0^{\delta t} + \int_{\delta t}^{[t]} + \int_{[t]}^t = I(t) + J(t) + K(t). \tag{3.10}$$

3. First we will prove that

$$\lim_{t \to 0} \frac{I(t)}{z(t)U(t)} = \beta \int_0^\delta (1 - u)^{-\gamma} u^{\gamma - 1} du. \tag{3.11}$$

Indeed

$$\frac{I(t)}{z(t)U(t)} = \frac{1}{z(t)U(t)} \int_0^{t\delta} z(t - u) dU(u)$$

$$= \int_0^{t\delta} \frac{z(t - u)}{z(t)} \frac{dU(u)}{U(t)}$$

$$= \int_0^\delta \frac{z(t(1 - x))}{z(t)} \frac{d_x U(tx)}{U(t)}. \tag{3.12}$$

Since $f_t(x) = z(t(1 - x))/z(t) \to (1 - x)^{-\gamma}$, $t \to \infty$ for every $x \in [0, \delta]$, $f_t(x)$ is monotone increasing as a function of x and the interval $[0, \delta]$ is closed, the convergence is uniform. On the other hand, from Erickson's theorem, it follows that $d_x U(tx)/U(t) \to \beta x^{\beta - 1}$, $t \to \infty$. Passing to the limit under the integral in (3.12) we obtain (3.11).

4. Now we will prove that

$$\limsup_{t \to \infty} \frac{J(t)\mu(t)}{\int_0^t z(u) du} \le c(1 - \delta)^{1 - \gamma}, \tag{3.13}$$

where c is a constant independent of δ and ε. First of all

$$J(t)\mu(t) = \mu(t) \int_{\delta t}^{[t]} z(u)dU(u) \leq \mu(t) \int_{[\delta t]}^{[t]} z(t-u)dU(u)$$

$$\leq \mu(t) \sum_{k=[\delta t]}^{[t]-1} z(t-k-1)(U(k+1)-U(k))$$

$$\leq (\mu(t)/\mu([\delta t])) \sum_{k=[\delta t]}^{[t]-1} z(t-k-1)\mu(k)(U(k+1)-U(k)). \quad (3.14)$$

Having in mind (3.7) we obtain for t large enough that

$$\mu(k)(U(k+1)-U(k)) \leq C_\beta + 1,$$

for all $k \in [[\delta t], [t]]$. So, we obtain from (3.14) that

$$J(t)\mu(t) \leq (\mu(t)/\mu([\delta t]))(C_\beta+1) \sum_{k=[\delta t]}^{[t]-1} z(t-k-1)$$

$$\leq (\mu(t)/\mu([\delta t]))(C_\beta+1) \left(\int_{t-[t]}^{t-[\delta t]-1} z(u)du + z(t-[t]) \right)$$

$$\leq (\mu(t)/\mu([\delta t]))(C_\beta+1) \left(\int_0^{t(1-\delta)} z(u)du + z(0) \right).$$

Now from (3.7) and (3.8) we get for the right-hand side

$$\limsup_{t\to\infty}(\mu(t)/\mu([\delta t]))(C_\beta+1) \left(\int_0^{t(1-\delta)} z(u)du + z(0) \right)$$

$$\leq \delta^{1-\beta}(C_\beta+1)(1-\delta)^{1-\gamma},$$

which proves (3.13).

5. Let us prove that

$$\limsup_{t\to\infty} \frac{K(t)\mu(t)}{\int_0^t z(u)du} = 0. \quad (3.15)$$

Indeed,

$$\frac{\mu(t)K(t)}{\int\limits_0^t z(u)du} \leq \frac{\mu(t)}{\int\limits_0^t z(u)du} \int\limits_{[t]}^t z(t-u)dU(u)$$

$$\leq \frac{\mu(t)}{\int\limits_0^t z(u)du} z(0)(U(t) - U([t]))$$

$$\leq \frac{\mu(t)}{\int\limits_0^t z(u)du} (U(t) - U(t-1)).$$

Now the assertion (3.15) follows from (3.7) and (3.8).

6. Finally, we obtain

$$\left| \frac{U * z(t)}{z(t)U(t)} - \beta B(\beta, 1 - \gamma) \right|$$

$$\leq \left| \frac{I(t)}{z(t)U(t)} - \beta \int\limits_0^\delta (1-u)^{-\gamma} u^{\beta-1} du \right|$$

$$+ \left| \frac{J(t)}{z(t)U(t)} \right| + \left| \frac{K(t)}{z(t)U(t)} \right|$$

$$+ \left| \beta B(\beta, 1 - \gamma) - \beta \int\limits_0^\delta (1-u)^{-\gamma} u^{1-\beta} du \right|.$$

Using (3.9), (3.10), (3.11), (3.13), and (3.15) we get

$$\limsup_{t \to \infty} \left| \frac{U * z(t)}{z(t)U(t)} - \beta B(\beta, 1 - \gamma) \right| \leq (c + 1)\varepsilon.$$

Since $\varepsilon > 0$ was arbitrary we complete the proof of the theorem. △

Remark 3.2 Similar results hold for the renewal function $U^d(t)$ of an delayed renewal process. Suppose that $S_n, n = 0, 1, 2, \ldots$ is a delayed renewal process, and $S_0 = T_1$ has a proper distribution function $G(t)$. We have already seen that $U^d(t) = U * G(t)$, that is $U^d(t)$ satisfies the renewal equation

$$U^d(t) = G(t) + \int\limits_0^t U^d(t-u)dF(u).$$

The elementary renewal theorem for $U^d(t)$ follows directly from the monotony of $U^d(t)$, the representation

$$\hat{U}^d(s) = \hat{G}(s)\hat{U}(s), \quad s > 0$$

of its Laplace transform, and Karamata's Tauberian theorem for Laplace transforms (Theorem A.4).

Theorem 3.7 *Suppose the conditions (a) and (b) hold. If $G(t)$ is a proper distribution function then $U^d(t)$ is regularly varying with exponent β and*

$$U^d(t) \sim C_\beta \frac{t}{\mu(t)}, \quad as \ t \to \infty.$$

Theorem 3.8 *Assume the conditions (a) and (b) with $\beta \in (\frac{1}{2}, 1]$. If $G(t) - G(t-h) = O(1/t)$ and $G(t) - G(t-h)$ is dRi for any fixed $h > 0$, then*

$$\lim_{t \to \infty} \mu(t)[U^d(t) - U^d(t-h)] = C_\beta h.$$

The proof follows immediately from Theorem 3.5. The conditions on $G(t)$ seems to be very restrictive, but they are satisfied at least in the following important cases:

1. $E[T_1] < \infty$. Because in this case for every fixed $h > 0$

$$0 \le G(t) - G(t-h) \le 1 - G(t-h), \quad t \ge 0,$$
$$1 - G(t-h) \downarrow 0 \text{ as } t \to \infty, \text{ and}$$
$$\int_0^\infty (1 - G(t-h))dt = h + E[T_1] < \infty.$$

2. T_1 has a Pareto distribution $G(t) = 1 - (a/t)^\beta, t \ge a > 0, \beta > 0$. In this case, for any fixed $h > 0$ and $t \ge a$

$$G(t) - G(t-h) = \left(\frac{a}{t-h}\right)^\beta - \left(\frac{a}{t}\right)^\beta = O\left(\frac{1}{t^{\beta+1}}\right),$$

which yields that $G(t) - G(t-h)$ is integrable on $[a, \infty)$. On the other hand, it is not difficult to be seen that it is decreasing. Thus, the function is dRi.

The asymptotic behavior of $Z(t) = U^d * z(t)$ follows from the associative property of the convolution $Z(t) = U^d * z(t) = (U * G) * z(t) = U * (G * z)(t)$.

Theorem 3.9 *Assume the conditions of Theorem 3.5. If $G(t)$ is proper, then*

$$\lim_{t \to \infty} \mu(t)[U^d * z(t)] = C_\beta \int_0^\infty z(t)dt.$$

Proof If $z(t)$ is dRi so is $G * z(t)$ since $G * z(t) \leq z(t)$. Now, we can apply Theorem 3.5 to the convolution $U * (G * z)(t)$ to get

$$\lim_{t \to \infty} \mu(t)[U * (G * z)(t)]$$

$$= C_\beta \int_0^\infty (G * z)(t)dt = C_\beta \int_0^\infty \left(\int_0^t z(t-u)dG(u) \right) dt$$

$$= C_\beta \left(\int_0^\infty z(t)dt \right) \left(\int_0^\infty dG(t) \right) = C_\beta \int_0^\infty z(t)dt. \qquad \triangle$$

Theorem 3.10 *Assume the conditions of Theorem 3.5. If $G(t)$ is a proper distribution function, then*

$$U^d * z(t) \sim C(\beta, \gamma) \frac{\int_0^t z(u)du}{\mu(t)}, \quad t \to \infty,$$

Proof Under the conditions of the theorem it is not difficult to be seen that $G * z(t)$ is nonincreasing and $G * z(t) \sim z(t)$ as $t \to \infty$. Now an application of Theorem 3.6 to the convolution $U * (G * z)(t)$ completes the proof. $\qquad \triangle$

3.1.4 Limit Theorems for Lifetimes. Dynkin–Lamperti Theorem

Equations (3.1), (3.3) and (3.4) are equivalent forms of the condition for the lifetime processes $A(t)$ and $B(t)$ to have a linear growth to infinity, i.e., the normalized processes $A(t)/t$ and $B(t)/t$ to have nondegenerate limit laws, jointly or separately. This result is known as Dynkin–Lamperti theorem. It is proved by Dynkin [12] and Lamperti [23].

Theorem 3.11 (Dynkin–Lamperti Theorem) *The condition (3.1), (3.3) and (3.4) with $\beta \in (0, 1)$ is necessary and sufficient for the existence of a nondegenerate limit law as $t \to \infty$ for each of $A(t)/t$, $B(t)/t$, and $(A(t)/t, B(t)/t)$. The corresponding limit laws are*

$$\lim_{t\to\infty} \Pr\left\{\frac{A(t)}{t} \le x\right\} = B_x(1-\beta,\beta), \quad 0 < x < 1,$$

$$\lim_{t\to\infty} \Pr\left\{\frac{B(t)}{t} \le x\right\} = \frac{\sin \pi\beta}{\pi} \int_0^x u^{-\beta}(1+u)^{-1}du, \quad x > 0,$$

$$\lim_{t\to\infty} \Pr\left\{\frac{A(t)}{t} \le x, \frac{B(t)}{t} \le y\right\}$$

$$= \frac{\beta \sin \pi\beta}{\pi} \int_0^x \int_0^y (1-u)^{\beta-1}(u+v)^{-\beta-1}dudv,$$

$$0 < x < 1, \quad y > 0.$$

Proof Let us prove the sufficiency of the first assertion. The one-dimensional distribution of the spent lifetime $A(t), t \ge 0$ satisfies the equation (see Theorem 1.17, A.)

$$\Pr\{A(t) \le x\} = (1-F(t))1_{[0,x]}(t) + \int_0^t \Pr\{A(t-u) \le x\}dF(u).$$

The solution of this equation is

$$\Pr\{A(t) \le x\} = \int_0^t (1-F(t-u))1_{[0,x]}(t-u)dU(u) = \int_{t-x}^t (1-F(t-u))dU(u).$$

Assume that $0 < x_1 < x_2 < 1$. Then

$$\Pr\{1-x_2 < \frac{A(t)}{t} \le 1-x_1\} = \Pr\{t(1-x_1) < A(t) \le t(1-x_1)\}$$

$$= \Pr\{A(t) \le t(1-x_1)\} - \Pr\{A(t)(t) \le t(1-x_2)\}$$

$$= \int_{t-t(1-x_1)}^t (1-F(t-u)dU(u) - \int_{t-t(1-x_2)}^t (1-F(t-u)dU(u)$$

$$= \int_{tx_1}^t (1-F(t-u)dU(u) - \int_{tx_2}^t (1-F(t-u)dU(u)$$

$$= \int_{tx_1}^{tx_2} (1-F(t-u)dU(u)$$

[Changing the variables by $u = vt$ to get]

$$\int\limits_{x_1}^{x_2} (1 - F(t(1 - v)))d_v U(tv)$$

$$= (1 - F(t))U(t) \int\limits_{x_1}^{x_2} \frac{1 - F(t(1 - v))}{1 - F(t)} \frac{d_v U(tv)}{U(t)}.$$

In a similar way as in the proof of the Key renewal theorem (Theorem 3.6), we can conclude that

$$\int\limits_{x_1}^{x_2} \frac{1 - F(t(1 - v))}{1 - F(t)} \frac{d_v U(tv)}{U(t)} \to \beta \int\limits_{x_1}^{x_2} (1 - v)^{-\beta} v^{\beta-1} dv.$$

Moreover,

$$(1 - F(t)) U(t) \to \frac{1}{\Gamma(1 - \beta)\Gamma(1 + \beta)} = \frac{1}{\Gamma(1 - \beta)\beta\Gamma(\beta)} = \frac{\sin \pi \beta}{\pi \beta}, \ t \to \infty.$$

In this way we have proved that for $0 < x < 1$, $\Pr\{A(t)/t \le x\}$ converges to the distribution function with density

$$\frac{\sin \pi \beta}{\pi} (1 - x)^{-\beta} x^{\beta-1}.$$

The sufficiency of the other two limiting distributions are similar. The necessity is more complicated. It can be found for example in Bingham et al. [9].   △

The case when $\beta = 1$ differs essentially from that in the above theorem. In this case the linear normalization for $A(t)$ and $B(t)$ is stronger than necessary. Recall that in this case $\mu(t)$ is slowly varying at infinity. In particular, if $\mu < \infty$ then $\mu(t) \to \mu < \infty$ also varies slowly and no normalization needs. The following theorem completes the picture about the limiting distributions of the lifetime processes. The proof can be found in Erickson [13].

Theorem 3.12 *Assume that $F(t)$ satisfies the conditions (a), (b), with $\beta = 1$ and (3.2). Then for $x \in [0, 1]$ and $y \ge 0$*

$$\lim_{t \to \infty} \Pr \left\{ \frac{\mu(A(t))}{\mu(t)} \le x, \frac{\mu(B(t))}{\mu(t)} \le y \right\} = \min\{x, y\},$$

and for $x \in (0, 1)$

$$\lim_{t \to \infty} \Pr \left\{ \frac{\mu(A(t))}{\mu(t)} \le x \right\} = \lim_{t \to \infty} \Pr \left\{ \frac{\mu(B(t))}{\mu(t)} \le x \right\} = x.$$

3.1.5 Discrete Time Case

Let F be a lattice distribution on $(0, \infty)$ which we suppose, without loss of generality, has span 1. The renewal measure U defined by (2.1) is also lattice with span 1. Denote as in Chap. 2 p_t and u_t the mass assigned to the integer t by F and U. If p_t satisfies the following relation

$$1 - F(t) = \sum_{k=t+1}^{\infty} p_k \sim t^{-\beta} L(t), t \to \infty,$$

(equivalent to (3.1)) for some $0 \leq \beta \leq 1$ and slowly varying function L, then Garsia and Lamperti [16] proved that for $\frac{1}{2} < \beta < 1$,

$$\lim_{t \to \infty} u_t t^{1-\beta} L(t) = \frac{\sin \pi \beta}{\pi},$$

while for $0 < \beta \leq \frac{1}{2}$ the lim must be replaced by lim inf.

Erickson [13] mentioned that the relation does hold when $0 < \beta \leq \frac{1}{2}$ provided the limit is taken excluding a set of integers having density 0. He also proved that in the case $\beta = 1$,

$$\lim_{t \to \infty} \mu(t) u_t = 1,$$

where, as before,

$$\mu(t) = \int_0^t (1 - F(x)) dx = \sum_{k=1}^{t} \left(\sum_{j=k}^{\infty} p_j \right).$$

3.2 Alternating Renewal Processes

The alternating renewal processes can be defined by a sequence of i.i.d. nonnegative random vectors (R_i, W_i) with independent coordinates, where R_i is interpreted as a *repairing* period (waiting time, OFF-period) of the ith element and W_i is the *work* period (or work time, or ON-period) which follows.

Denote by $F_W(t) := \Pr\{W_i \leq t\}$, $F_R(t) := Pr\{R_i \leq t\}$, and $T_i = R_i + W_i$, $i = 1, 2, \ldots$, the sequence of independent, identically distributed, nonnegative r.v. with d.f. $F(t) := \Pr\{T_i \leq t\} = (F_R * F_W)(t)$.

Two types of renewal epochs appear: $S_n = \sum_{i=1}^{n} (R_i + W_i)$, $n \geq 0$, the end of the nth lifetime which coincides with the beginning of the next waiting period, and $S'_{n+1} = S_n + R_{n+1}$, $n \geq 0$, the end of the $(n + 1)$st waiting period which coincides with the beginning of the next lifetime.

An ordinary renewal process $N(t) = \max\{n : S_n \le t\}$ is defined by the sequence $S_0 = 0$, $S_{n+1} = S_n + T_{n+1}$, $n = 0, 1, 2, \ldots$. Let us define the following process

$$\sigma(t) = t - S'_{N(t)+1} = t - S_{N(t)} - R_{N(t)+1}, \quad t \ge 0.$$

Evidently, if all R_i are identically equal to zero we have an ordinary renewal process. In this case, $\sigma(t) \equiv A(t)$ is always nonnegative. For an alternating renewal process $\sigma(t)$ can take both negative and positive values

$$\sigma(t) = \sigma^+(t) - \sigma^-(t),$$

where $\sigma^+(t) = \max\{\sigma(t), 0\}$ and $\sigma^-(t) = \max\{-\sigma(t), 0\}$. Note that $\sigma^+(t)$ can be interpreted as the spent work time and $\sigma^-(t)$ can be considered as the residual waiting period. Therefore, if for a given t, $\sigma(t) \ge 0$ then the process is in its ON-period, but if $\sigma(t) < 0$ then the process is in its OFF-period.

An interesting characteristic of the alternating renewal processes is the probability

$$\Pr\{\sigma(t) \ge 0\} \equiv \Pr\{\text{the process is in ON period at time } t\}$$

or its limiting behavior as $t \to \infty$. It appears that the limiting behavior depends essentially on the mathematical expectations of the lengths of ON-and OFF-periods. The following theorems clarify the picture.

Assume the following basic conditions:

$$F_R(0) = F_W(0) = 0, \quad F_R(x) \text{ and } F_W(x) \text{ are nonlattice d.f.}$$

(R.1) $m_R = \mathbf{E}[R_i] < \infty$.

(R.2) $\mathbf{E}[R_i] = \infty$, $1 - F_R(t) \sim t^{-\alpha} L_R(t)$, $t \to \infty$, $\alpha \in (\frac{1}{2}, 1]$, $L_R(.)$ is a s.v.f., and for each $h > 0$ fixed $A(t) - A(t - h) = O(1/t)$, $t > 0$.

(W.1) $m_W = \mathbf{E}[W_i] < \infty$.

(W.2) $\mathbf{E}[W_i] = \infty$, $1 - F_W(t) \sim t^{-\beta} L_W(t)$, $t \to \infty$, $\beta \in (\frac{1}{2}, 1]$, where $L_W(.)$ is a s.v.f.

(RW) The following limit exists $\lim_{t \to \infty}(1 - F_R(t))/(1 - F_W(t)) = c$, $0 \le c \le \infty$.
Let us denote also

$$m_R(t) = \int_0^t (1 - F_R(x))dx, \ m_W(t) = \int_0^t (1 - F_W(x))dx, \ \mu(t) = \int_0^t (1 - F(x))dx.$$

Theorem 3.13 *Assume (R.1) and (W.1). Then*

$$\lim_{t \to \infty} \Pr\{\sigma(t) \geq 0\} = \frac{m_W}{m_R + m_W}.$$

Theorem 3.14 *Assume (R.2), $c = \infty$ and either (W.1) or (W.2) is fulfilled. Then*

$$\lim_{t \to \infty} \Pr\{\sigma(t) \geq 0\} = 0.$$

(i) *If only (W.1) holds then*

$$\Pr\{\sigma(t) \geq 0\} \sim L_1(t)t^{-(1-\alpha)},$$

where $L_1(t)$ is a s.v.f. as $t \to \infty$.
  (ii) *If only (W.2) with $1/2 < \beta < 1$ holds then*

$$\Pr\{\sigma(t) \geq 0\} \sim L_2(t)t^{-(\beta-\alpha)},$$

where $L_2(t)$ is a s.v.f., as $t \to \infty$.
  (iii) *If only (W.2) with $\beta = 1$ holds then*

$$\Pr\{\sigma(t) \geq 0\} \sim L_3(t)t^{-(1-\alpha)},$$

where $L_3(t)$ is a s.v.f. as $t \to \infty$.

Theorem 3.15 *Assume (R.2), $c = \infty$ and either (W.1) or (W.2) is fulfilled.*
  (i) *If only (W.1) holds then for $x \geq 0$*

$$\lim_{t \to \infty} \Pr\{\sigma(t) \leq x | \sigma(t) \geq 0\} = \frac{m_W(x)}{m_W}.$$

(ii) *If only (W.2) with $1/2 < \beta < 1$ holds then for $0 \leq x \leq 1$*

$$\lim_{t \to \infty} \Pr\left\{\frac{\sigma(t)}{t} \leq x | \sigma(t) \geq 0\right\} = \frac{B_x\alpha, 1 - \beta)}{B(\alpha, 1 - \beta)}.$$

(iii) *If only (W.2) with $\beta = 1$ holds then for $0 \leq x \leq 1$*

$$\lim_{t \to \infty} \Pr\left\{\frac{m_W(\sigma(t))}{m_W(t)} \leq x | \sigma(t) \geq 0\right\} = x.$$

Theorem 3.16 *Assume (W.2), $0 \leq c < \infty$ and either (R.1) or (R.2) is fulfilled. Then*

$$\lim_{t \to \infty} \Pr\{\sigma(t) \geq 0\} = \frac{1}{1 + c}.$$

(i) If (W.2) with $1/2 < \beta < 1$ holds then for $0 \leq x \leq 1$

$$\lim_{t \to \infty} \Pr \left\{ \frac{\sigma(t)}{t} \leq x \right\} = \frac{c}{1+c} + \frac{1}{1+c} \frac{B_x(\beta, 1-\beta)}{B(\beta, 1-\beta)}.$$

and

$$\lim_{t \to \infty} \Pr \left\{ \frac{\sigma(t)}{t} \leq x \mid \sigma(t) \geq 0 \right\} = \frac{B_x(\beta, 1-\beta)}{B(\beta, 1-\beta)}.$$

(ii) If (W.2) with $\beta = 1$ holds then for $0 \leq x \leq 1$

$$\lim_{t \to \infty} \Pr \left\{ \frac{m_F(\sigma(t))}{m_F(t)} \leq x \right\} = \frac{c}{1+c} + \frac{x}{1+c}$$

and

$$\lim_{t \to \infty} \Pr \left\{ \frac{m_F(\sigma(t))}{m_F(t)} \leq x \mid \sigma(t) \geq 0 \right\} = x.$$

The proofs of these theorems can be found in Feller [14] (Theorem 3.13) and Mitov and Yanev [25] (Theorems 3.14–3.16).

3.3 Renewal Reward Processes

3.3.1 Bivariate Renewal Process

Consider a sequence of i.i.d. random vectors

$$(T^{(1)}, T^{(2)}), (T_1^{(1)}, T_1^{(2)}), \ldots, (T_n^{(1)}, T_n^{(2)}), \ldots$$

with common distribution function $F(x, y)$ and marginal d.f. $F_1(x)$, $F_2(x)$. As before we set $S_0^{(1)} = S_0^{(2)} = 0$ and $S_n^{(1)} = \sum_{i=1}^{n} T_i^{(1)}$, $S_n^{(2)} = \sum_{i=1}^{n} T_i^{(2)}$. Starting from the sequences of partial sums we can consider several counting processes:

$$N^{(1)}(x) = \max \left\{ n \geq 1 : S_n^{(1)} \leq x \right\}, \quad N^{(2)}(y) = \max \left\{ n \geq 1 : S_n^{(2)} \leq y \right\},$$

$$M^{(1)}(x) = \min \left\{ n \geq 1 : S_n^{(1)} > x \right\}, \quad M^{(2)}(y) = \min \left\{ n \geq 1 : S_n^{(2)} > y \right\}.$$

A bivariate renewal counting process can be defined as follows:

$$N(x, y) = \max \left\{ n : S_n^{(1)} \leq x, \ S_n^{(2)} \leq y \right\}$$

or as

$$M(x, y) = \min \left\{ n : S_n^{(1)} > x \ \text{ or } \ S_n^{(2)} > y \right\}.$$

With $N(x, y)$ we count the number of terms that are in the region $(0, x] \times (0, y]$ and with $M(x, y)$ we determine the time when we leave the region $(0, x] \times (0, y]$ for the first time. Clearly, we have $M(x, y) = \min \left\{ M^{(1)}(x), \ M^{(2)}(y) \right\}$ and

$$\Pr\{M(x, y) \le n\} = 1 - F^{n*}(x, y).$$

where

$$F^{1*}(x, y) = F(x, y),$$

$$F^{n*}(x, y) = \int_0^x \int_0^y F^{(n-1)*}(x - u, y - v) dF(u, v), \ n = 2, 3, \ldots.$$

The bivariate renewal function is given by $U(x, y) = \mathbf{E}[M(x, y)]$ and as in the univariate case, we have

$$U(x, y) = \sum_{n=0}^{\infty} F^{n*}(x, y).$$

If both means $\mu = \mathbf{E}[T^{(1)}] < \infty$, $\nu = \mathbf{E}[T^{(2)}] < \infty$ we have the following limit theorem (cf. Hunter [20, 21]).

Theorem 3.17 *As* $\min \{x, y\} \to \infty$, *we have:*

(i) $M(\mu x, \nu y) / \min\{x, y\} \overset{a.s.}{\longrightarrow} 1;$

(ii) $U(\mu x, \nu y) / \min\{x, y\} \longrightarrow 1.$

Proof (i) For $\min\{x, y\} > x^\circ$, we have

$$1 - \varepsilon \le M^{(1)}(\mu x)/x \le 1 + \varepsilon \text{ and } 1 - \varepsilon \le M^{(2)}(\nu y)/y \le 1 + \varepsilon.$$

First suppose that $x < y$. Then

$$M^{(1)}(\mu x) \ge x(1 - \varepsilon), \quad M^{(2)}(\nu y) \ge y(1 - \varepsilon) > x(1 - \varepsilon)$$

and

$$\min \left\{ M^{(1)}(\mu x), \ M^{(2)}(\nu y) \right\} \ge \min\{x, y\}(1 - \varepsilon).$$

On the other hand, we have

$$\min \left\{ M^{(1)}(\mu x), \ M^{(2)}(\nu y) \right\} \le M^{(1)}(\mu x) \le x(1 + \varepsilon) = \min\{x, y\}(1 + \varepsilon).$$

It follows that

$$\min\{x, y\}(1 - \varepsilon) \leq \min\left\{M^{(1)}(\mu x),\ M^{(2)}(\nu y)\right\} \leq \min\{x, y\}(1 + \varepsilon).$$

The first result follows.

(ii) For $U(x\mu, y\nu) = \mathbf{E}[M(x\mu, y\nu)]$, Fatou's lemma shows that

$$\liminf_{\min\{x,y\}\to\infty} \frac{U(x\mu, y\nu)}{\min\{x, y\}} \geq 1.$$

On the other hand, we have $M(x, y) \leq M^{(1)}(x)$, and $M(x, y) \leq M^{(2)}(y)$. It follows that $U(x, y) \leq U^{(1)}(x)$, and $U(x, y) \leq U^{(2)}(y)$. Using the Elementary renewal theorem, we can find x° so that $U(\mu x, \nu y) \leq x(1 + \varepsilon)$ and $U(\mu x, \nu y) \leq y(1 + \varepsilon)$ for $\min\{x, y\} \geq x^\circ$. We conclude that

$$\frac{U(x\mu, y\nu)}{\min\{x, y\}} \leq 1 + \varepsilon.$$

The result follows. $\triangle$

Remark 3.3 Bickel and Yahav [8] also proved a Blackwell type of result. Among others they proved that

$$U(t + a, t + a) - U(t, t) \to a \min\left\{\frac{1}{\mu}, \frac{1}{\nu}\right\}.$$

3.3.2 Renewal Reward Processes

Let us consider a sequence of i.i.d. nonnegative random vectors

$$(T, Y), (T_1, Y_1), (T_2, Y_2), \ldots$$

with common distribution function $F(x, y)$ and marginal d.f. $F_1(x)$ and $F_2(x)$. By the sequence $\{T_i\}$ we define the renewal sequence

$$S_0 = 0,\ S_n = \sum_{i=1}^{n} T_i$$

and the corresponding renewal counting processes $N(t)$ and $M(t)$. Define also

$$S_Y(0) = 0,\ \text{and}\ S_Y(n) = \sum_{i=1}^{n} Y_i,$$

We can use the following interpretation. We have renewals with interarrival times T_i. Suppose that a reward (or fine) Y_n is earned during the nth renewal. The total reward earned by the nth renewal is given by $S_Y(n)$. The total reward earned by time t is given by $Y(t) = S_Y(M(t))$. The d.f. of $Y(t)$ is rather complicated but we have the following result.

Proposition 3.1 *The distribution function of $Y(t)$ is given by*

$$\Pr\{Y(t) > x\} = \int_0^t \int_0^x \Pr\{Y > x - v\}\, dU(u, v),$$

where $U(x, y) = \sum_{n=0}^{\infty} F^{n*}(x, y)$.

Proof By conditioning on $M(t)$, we have

$$\Pr\{Y(t) \le x\} = \sum_{n=1}^{\infty} \Pr\{S_Y(n) \le x, M(t) = n\}$$

$$= \sum_{n=1}^{\infty} \Pr\{S_Y(n) \le x, S_{n-1} \le t < S_n\}.$$

By conditioning on $\{S_{n-1}, S_Y(n-1)\}$, we get that

$$\Pr\{Y(t) \le x\} = \sum_{n=1}^{\infty} \int_{u=0}^{\infty} \int_{v=0}^{\infty} \Pr\{Y_n \le x - v, u \le t < u + T_n\}\, dF^{(n-1)*}(u, v)$$

$$= \int_{u=0}^{t} \int_{v=0}^{x} \Pr\{Y \le x - v, t - u < T\}\, dU(u, v).$$

Rearranging terms, we find that

$$\Pr\{Y(t) \le x\} = \int_{u=0}^{t} \int_{v=0}^{x} \Pr\{Y \le x - v\} - \Pr\{Y \le x - v, T \le t - u\}\, dU(u, v)$$

$$= \int_{u=0}^{t} \int_{v=0}^{x} \Pr\{Y \le x - v\}\, dU(u, v) - (F * U)(t, x)$$

$$= \int_{u=0}^{t} \int_{v=0}^{x} \Pr\{Y \le x - v\}\, dU(u, v) - U(t, x) + 1$$

and the result follows.   △

For the mean value of $Y(t)$ we have the following result.

Corollary 3.1   *The mean* $\mathbf{E}[Y(t)]$ *is given by*

$$\mathbf{E}[Y(t)] = \nu U(t),$$

where $\nu = \mathbf{E}[Y]$ *and* $U(t) = \mathbf{E}[M(t)]$ *is the renewal function.*

Proof We have

$$\mathbf{E}[Y(t)] = \int_{x=0}^{\infty} \int_{u=0}^{t} \int_{v=0}^{x} \Pr\{Y > x - v\} dU(u, v) dx$$

$$= \int_{u=0}^{t} \int_{v=0}^{\infty} \int_{x=v}^{\infty} \Pr\{Y > x - v\} dU(u, v) dx.$$

It follows that

$$\mathbf{E}[Y(t)] = \nu \int_{u=0}^{t} \int_{v=0}^{\infty} dU(u, v) = \nu U(t). \qquad \triangle$$

The next theorem provides the asymptotic behavior of the average earnings.

Theorem 3.18   *Suppose that* $\mu = \mathbf{E}[T] < \infty$ *and* $\nu = \mathbf{E}[Y] < \infty$. *Then*
(i) $Y(t)/t \xrightarrow{P} \nu/\mu$;
(ii) $\mathbf{E}[Y(t)]/t \longrightarrow \nu/\mu$;
(iii) *In the nonlattice case, we have* $\mathbf{E}[Y(t + h)] - \mathbf{E}[Y(t)] \to h\nu/\mu$.

Proof (i) We write
$$\frac{Y(t)}{t} = \frac{S_Y(M(t))}{M(t)} \times \frac{M(t)}{t}.$$

Since $M(t) \xrightarrow{a.s.} \infty$ and $M(t)/t \xrightarrow{a.s.} 1/\mu$, we find the first result.

(ii) To prove the second result, note that $M(t)$ is a stopping time for $Y_1, Y_2, \ldots$. To see this, note that the independence of $\{M(t) = n\}$ and $\{T_{n+1}, T_{n+2}, \ldots\}$ implies the independence of $\{M(t) = n\}$ and $\{Y_{n+1}, Y_{n+2}, \ldots\}$. Using Wald's identity (Lemma 1.2), we get that $\mathbf{E}[Y(t)] = \mathbf{E}[M(t)]\mathbf{E}[Y] = \nu U(t)$. Now the result follows.

(iii) Use Blackwell's renewal theorem in the previous expression. $\triangle$

Remark 3.4 Often the total reward is given by $R(t) = S_Y(N(t))$, where $N(t) = M(t) - 1$. It is clear that $Y(t) - R(t) = Y_{M(t)}$.

Proposition 3.2 *We have*

$$\Pr\{Y_{M(t)} > x\} = \int_0^t \Pr\{T > t - u, Y > x\} dU(u),$$

and

$$\lim_{t \to \infty} \Pr\{Y_{M(t)} > x\} = \frac{1}{\mu} \int_0^\infty \Pr\{T > u, Y > x\} du.$$

Proof By conditioning on $M(t)$, we have

$$\Pr\{Y_{M(t)} \leq x\} = \sum_{n=1}^\infty \Pr\{Y_n \leq x, \ S_{n-1} \leq t < S_n\}$$

$$= \sum_{n=1}^\infty \Pr\{Y_n \leq x, \ S_{n-1} \leq t < S_{n-1} + T_n\}.$$

It follows that

$$\Pr\{Y_{M(t)} > x\} = \int_0^t \sum_{n=1}^\infty \Pr\{Y_n > x, u \leq t < u + T_n\} dF_1^{(n-1)*}(u)$$

$$= \int_0^t \Pr\{Y > x, t - u < T\} dU(u),$$

This proves the first result. Since $\Pr\{T > t - u, Y > x\} \leq \Pr\{T > t - u\}$, the Key renewal theorem leads to

$$\lim_{t \to \infty} \Pr\{Y_{M(t)} > x\} = \frac{1}{\mu} \int_0^\infty \Pr\{T > u, Y > x\} du,$$

which is the second result. △

Remark 3.5 Note that this result is similar to the result for $T_{N(t)}$ (see Theorem 1.18), where we proved that

$$\Pr\{T_{N(t)} \leq x\} \to \frac{1}{\mu} \int_0^x u \, dF_0(u).$$

Moreover, the following result holds true.

Proposition 3.3 *As $t \to \infty$, we have $t^{-1}\mathbf{E}[Y_{M(t)}] \to 0$.*

Proof Note that

$$\mathbf{E}[Y_{M(t)}] = \int_{x=0}^{\infty} \int_{0}^{t} \Pr\{T > t - u, Y > x\}dU(u)dx$$

$$= \int_{u=0}^{t} \int_{x=0}^{\infty} \Pr\{T > t - u, Y > x\}dxdU(u)$$

$$= \int_{u=0}^{t} h(t - u)dU(u),$$

where

$$h(t) = \int_{0}^{\infty} \Pr\{T > t, Y > x\}dx.$$

Since $\Pr\{T > t, Y > x\} \leq \Pr\{Y > x\}$, we obtain that $h(t) \leq v$. Moreover, as $t \to \infty$, we have $h(t) \longrightarrow 0$ by dominated convergence. Now choose $t^{\circ} > 0$ and $\varepsilon > 0$ so that

$$0 \leq h(t) \leq \varepsilon \text{ for } t \geq t^{\circ}.$$

With this choice, we have

$$\mathbf{E}[Y_{M(t)}] = \int_{0}^{t-t^{\circ}} h(t - u)dU(u) + \int_{t-t^{\circ}}^{t} h(t - u)dU(u) = J_1(t) + J_2(t).$$

For the first term we get $J_1(t) \leq \varepsilon U(t - t^{\circ})$, while for the second term we get that

$$J_2(t) \leq v(U(t) - U(t - t^{\circ})).$$

It follows that

$$0 \leq \limsup_{t \to \infty}(J_1(t) + J_2(t))/t \leq \varepsilon\mu.$$

Since ε was arbitrary, the result follows. $\triangle$

Corollary 3.2 *Suppose that $\mu < \infty$ and $v < \infty$. Then*
(i) $R(t)/t \xrightarrow{P} v/\mu$;
(ii) $\mathbf{E}[R(t)]/t \longrightarrow v/\mu$.

In the next result, we obtain a central limit theorem for $Y(t)$ (cf. e.g., Gut [19], Ch. IV, Theorem 2.3).

Remark 3.6 Recall that a vector (Z_1, Z_2) has a standard bivariate normal distribution with correlation coefficient parameter ρ if its density is given by

$$\varphi_\rho(x, y) = \frac{1}{2\pi\sqrt{1-\rho^2}} \exp(-(x^2 + y^2 - 2\rho xy)/2(1 - \rho^2)).$$

Theorem 3.19 *Suppose that (T, Y) has finite means (μ, ν), finite variances (σ_T^2, σ_Y^2) and correlation coefficient ρ. Then*

$$\frac{Y(t) - t\nu/\mu}{\sqrt{t}} \xrightarrow{d} \frac{1}{\mu\sqrt{\mu}}(-\nu\sigma_1 Z_1 + \mu\sigma_2 Z_2),$$

where the random vector (Z_1, Z_2) has the standard bivariate normal distribution with parameter ρ.

Proof From the usual central limit theorem, we have that

$$a\frac{S_n - n\mu}{\sqrt{n}} + b\frac{S_Y(n) - n\nu}{\sqrt{n}} \xrightarrow{d} a\sigma_1 Z_1 + b\sigma_2 Z_2.$$

By a theorem of Anscombe (see, e.g., Gut [19], Ch. I, Theorem 3.1), we can replace n by $M(t)$ and $\sqrt{n}$ by $\sqrt{t/\mu}$. Then we get that

$$a\frac{S_{M(t)} - M(t)\mu}{\sqrt{t}} + b\frac{S_Y(M(t)) - M(t)\nu}{\sqrt{t}} \xrightarrow{d} \frac{1}{\sqrt{\mu}}(a\sigma_1 Z_1 + b\sigma_2 Z_2).$$

From $S_{M(t)-1} \le t < S_{M(t)}$ we get

$$\frac{t - M(t)\mu}{\sqrt{t}} \le \frac{S_{M(t)} - M(t)\mu}{\sqrt{t}} \le \frac{t - M(t)\mu}{\sqrt{t}} + \frac{1}{\sqrt{t}}T_{M(t)}.$$

We know that (see Theorem 1.18)

$$\Pr\{T_{M(t)} \le x\} \to \frac{1}{\mu} \int_0^x u\, dF_1(u).$$

So,

$$\frac{1}{\sqrt{t}}T_{M(t)} \xrightarrow{P} 0.$$

Now we conclude that

$$a\frac{t - M(t)\mu}{\sqrt{t}} + b\frac{S_Y(M(t)) - M(t)\nu}{\sqrt{t}} \xrightarrow{d} \frac{1}{\sqrt{\mu}}(a\sigma_1 Z_1 + b\sigma_2 Z_2).$$

Choosing $a = -\nu$ and $b = \mu$, we get that

$$\frac{\mu Y(t) - \nu t}{\sqrt{t}} \xrightarrow{d} \frac{1}{\sqrt{\mu}}(-\nu \sigma_T Z_1 + \mu \sigma_Y Z_2),$$

and hence the result. $\triangle$

3.4 Superposition of Renewal Processes

3.4.1 Introduction

The problem for superposition of a number of renewal processes can be stated as follows. Suppose that there are a number of sources, at each of which events occur from time to time, and that the outputs of the sources are combined into one pooled output. More precisely, assume that a family

$$T_1^{(i)}, T_2^{(i)}, T_3^{(i)}, \ldots, T_n^{(i)}, \ldots, \quad i = 1, 2, \ldots, \tag{3.16}$$

is given. Denote the distribution functions as follows

$$G^{(i)}(t) = \Pr\{T_1^{(i)} \le t\}, \text{ and } F^{(i)}(t) = \Pr\{T_n^{(i)} \le t\}, \quad n = 2, 3, \ldots.$$

They define a family of independent renewal sequences and their counting renewal processes

$$\{S_n^{(i)}\}_{n=0}^{\infty}, \quad i = 1, 2, \ldots, \tag{3.17}$$

$$\{N^{(i)}(t), \ t \ge 0\}, \quad i = 1, 2, \ldots$$

Consider the sequence $Z_n^{(m)}$, $n = 1, 2, 3, \ldots$ of events obtained by combining the events from the first m sequences $\{S_n^{(i)}\}_{n=0}^{\infty}$, $i = 1, 2, \ldots, m$, and rearranging them in an increasing order. A natural interpretation of the sequence $Z_n^{(m)}$, $n = 0, 1, 2, \ldots$ can be done in terms of queueing theory. If one considers the sequence $S_n^{(i)}$, $n = 0, 1, 2, \ldots$ as the instants when customers come for service in a system from source $i, i = 1, 2, \ldots, m$ then the sequence $Z_n^{(m)}$ represents the instants when customers come into the system from any of the sources $1, 2, \ldots, m$. In this interpretation the sum

$$N_m(t) = \sum_{i=1}^{m} N^{(i)}(t)$$

represents the total number of customers coming into the system during the time interval $(0, t]$.

Let us denote the spent and residual lifetimes associated with the sequence $\{S_n^{(i)}\}_{n=1}^{\infty}$, $i = 1, 2, \ldots$ as

$$A^{(i)}(t) = t - S_{N^{(i)}(t)}^{(i)}, \quad t > 0,$$

and

$$B^{(i)}(t) = S_{N^{(i)}(t)+1}^{(i)} - t, \quad t > 0.$$

Definition 3.1 The spent and residual lifetimes associated with the combined sequence $Z_n^{(m)}$ are defined as follows

$$A(m, t) = \min\{A^{(1)}(t), A^{(2)}(t), \ldots, A^{(m)}(t)\}, \quad t > 0, \quad m = 1, 2, \ldots,$$

and

$$B(m, t) = \min\{B^{(1)}(t), B^{(2)}(t), \ldots, B^{(m)}(t)\}, \quad t > 0, \quad m = 1, 2, \ldots.$$

Some interesting problems arise in this situation. Whether the superposition of a finite (or infinite) number of ordinary renewal processes is again an ordinary (or delayed) renewal process? Is it possible to find the distributions of the process $N_m(t), t \geq 0$? What is the limiting behavior of the number of events in an interval of fixed length, $[t - h, t]$, or $[0, t]$, as $t \to \infty$? What happens with the lifetime processes in the combined sequence, etc? In this section, we present some answers to these problems.

Example 3.2 Assume that $G^{(i)}(t) = F^{(i)}(t) = 1 - \exp(-\lambda_i t)$ for some $\lambda_i > 0$. Then $N^{(i)}(t)$ is a Poisson process and the renewal function is $U^{(i)}(t) = \lambda_i t + 1$ (see Sect. 1.3.2). It is not difficult to prove that the combined process $N_m(t)$ is also Poisson. The interarrival times have d.f. $F_m(t) = 1 - \exp(-(\sum_{i=1}^{m} \lambda_i)t)$ and the renewal function is $U_m(t) = (\sum_{i=1}^{m} \lambda_i)t + 1$. In this case, the answers to all of the above problems are clear.

In general, the superposition of renewal processes is not a renewal process. The finite dimensional distributions of $N_m(t), t \geq 0$ cannot be found easily. Fortunately, in some cases it is possible to approximate the process $N_m(t), t \geq 0$ by a Poisson process. The first result in this direction has been proved by Cox and Smith [10]. They assumed that the interarrival times in all renewal processes in (3.17) have the same distribution function $F(t)$ and finite mean μ. Under these conditions they proved that the process $N_m(t)$ converges in distribution to a Poisson process as $m \to \infty$. The limiting distributions of the spend and residual lifetimes of the combined process are exponential in this case.

The situation differs essentially in the case when the interarrival times have heavy tail distributions. Assume that interarrival times given in (3.16) are identically distributed and

$$1 - F(t) \sim t^{-\alpha}L(t), \quad t \to \infty, \tag{3.18}$$

where $0 < \alpha < 2$ and $L(t)$ is a slowly varying function at infinity. In this case, T_i may have finite mean and infinite variance $(1 < \alpha < 2)$, or both the variance and mean infinite $(0 < \alpha < 1)$. Different types of limiting processes are obtained in this case (see, e.g., Gajgalas and Kaj [15] and the references therein).

In case when $\alpha \in (0, 1)$ (the interarrival times have infinite mean) the limiting distributions of the spent and residual lifetimes of the combined process are not exponential as it is seen from the following theorems.

Theorem 3.20 (Mitov and Yanev [26]) *Assume* (3.18) *with* $0 < \alpha < 1$, *and* $t \to \infty$ *and* $m \to \infty$ *simultaneously.*

(i) *If* $mt^{-(1-\alpha)} \to C$ *and* $0 < C < \infty$ *then for any* $x > 0$

$$\lim_{m,t \to \infty} \Pr\{A(m, t) \le x\} = 1 - \exp\left(-C\frac{\sin \pi \alpha}{\pi(1 - \alpha)}x^{1-\alpha}\right).$$

(ii) *If* $mt^{-(1-\alpha)} \sim \theta(t)$, *where* $\theta(t) \to 0$ *as* $t \to \infty$, *then for any* $x > 0$

$$\lim_{m,t \to \infty} \Pr\left\{\theta(t)^{1/(1-\alpha)} A(m, t) \le x\right\} = 1 - \exp\left(-\frac{\sin \pi \alpha}{\pi(1 - \alpha)}x^{1-\alpha}\right).$$

(iii) *If* $mt^{-(1-\alpha)} \sim \theta(t)$, *where* $\theta(t) \to \infty$ *as* $t \to \infty$, *then for any* $x > 0$

$$\lim_{m,t \to \infty} \Pr\left\{\theta(t)^{1/(1-\alpha)} A(m, t) \le x\right\} = 1 - \exp\left(-\frac{\sin \pi \alpha}{\pi(1 - \alpha)}x^{1-\alpha}\right).$$

Theorem 3.21 (Mitov and Yanev [26]) *Assume* (3.18) *with* $0 < \alpha < 1$, *and* $t \to \infty$ *and* $m \to \infty$ *simultaneously.*

(i) *If* $mt^{-(1-\alpha)} \to C$, $0 < C < \infty$ *then for any* $x > 0$

$$\lim_{m,t \to \infty} \Pr\{B(m, t) \le x\} = 1 - \exp\left(-C\frac{\sin \pi \alpha}{\pi(1 - \alpha)}x^{1-\alpha}\right).$$

(ii) *If* $mt^{-(1-\alpha)} \sim \theta(t) \to 0$ *then for any* $x > 0$

$$\lim_{m,t \to \infty} \Pr\left\{\theta(t)^{1/(1-\alpha)} B(m, t) \le x\right\} = 1 - \exp\left(-\frac{\sin \pi \alpha}{\pi(1 - \alpha)}x^{1-\alpha}\right).$$

(iii) *If* $mt^{-(1-\alpha)} \sim \theta(t) \to \infty$ *then for any* $x > 0$

$$\liminf_{m,t \to \infty} \Pr\left\{\theta(t)^{1/(1-\alpha)} B(m, t) \le x\right\} \ge 1 - \exp\left(-\frac{\sin \pi \alpha}{\pi(1 - \alpha)}x^{1-\alpha}\right).$$

We conclude this section with a general result proved by Grigelionis [18]. Let us have a rectangular array of renewal counting processes

$$N^{(1,1)}(t), N^{(1,2)}(t), \ldots, N^{(1,r)}(t), \ldots$$
$$N^{(2,1)}(t), N^{(2,2)}(t), \ldots, N^{(2,r)}(t), \ldots$$

$$\ldots$$

$$N^{(n,1)}(t), N^{(n,2)}(t), \ldots, N^{(n,r)}(t), \ldots,$$

where the processes in each row are independent. The corresponding renewal functions are denoted by

$$U^{(n,r)}(t) = \mathbf{E}[N^{(n,r)}(t)], \ t \geq 0.$$

Let $k_n, n \geq 0$ be a sequence of positive integers, such that $k_n \uparrow \infty, n \to \infty$. Then for every $n = 1, 2, \ldots$ we define the superposition of the first k_n counting processes in the nth row,

$$N_n(t) = \sum_{r=1}^{k_n} N^{(n,r)}(t), \ t \geq 0.$$

Assume that the process $N^{(n,r)}(t), t \geq 0$ is determined by the distribution function $G^{(n,r)}(t)$ of the first interarrival time and $F^{(n,r)}(t)$ the distribution function of the other interarrival times. Suppose that the following conditions hold true. For every fixed $t \geq 0$,

$$\lim_{n \to \infty} \max_{1 \leq r \leq k_n} G^{(n,r)}(t) = 0, \quad \lim_{n \to \infty} \max_{1 \leq r \leq k_n} F^{(n,r)}(t) = 0, \tag{3.19}$$

$$\lim_{n \to \infty} \sum_{r=1}^{k_n} U^{(n,r)}(t) = U(t) < \infty. \tag{3.20}$$

Theorem 3.22 (Grigelionis [18]) *Suppose that conditions* (3.19) *and* (3.20) *hold. Then the finite dimensional distributions of the process* $N_n(t)$, $t \geq 0$ *converge to the finite dimensional distributions of a time inhomogeneous Poisson process* $N(t)$, $t \geq 0$ *with intensity* $U(t)$.

The one-dimensional distributions of the process $N(t)$ are

$$\Pr\{N(t) = k\} = \frac{U(t)^k}{k!} e^{-U(t)}, \ t \geq 0.$$

Grigelionis also proved the rate of convergence of one-dimensional distributions.

3.4.2 Superposition of Two Dependent Renewal Processes

In this section, we discuss the superposition of two dependent renewal processes. Suppose that

$$(T^{(1)}, T^{(2)}), (T_1^{(1)}, T_1^{(2)}), \ldots, (T_n^{(1)}, T_n^{(2)}), \ldots$$

are i.i.d. nonnegative random vectors. We do not assume that the components of the vector $(T_n^{(1)}, T_n^{(2)})$ are independent. So, the renewal sequences $S_n^{(1)}, n = 0, 1, 2, \ldots$ and $S_n^{(2)}, n = 0, 1, 2, \ldots$ and the corresponding renewal counting processes $N^{(1)}(t)$, $t \geq 0$ and $N^{(2)}(t), t \geq 0$ are also not independent. The combined process $\{Z_n, n = 0, 1, 2, \ldots\}$ is not a renewal process in general, but we can still define

$$N(t) = \sup \{n : Z_n \leq t\}.$$

Clearly we have

$$N(t) = N^{(1)}(t) + N^{(2)}(t)$$

and we can transfer some properties from the ordinary counting processes to $N(t)$. Taking expectations, one obtains

$$U(t) = U^{(1)}(t) + U^{(2)}(t).$$

The next result follows from Theorems 1.9, 1.10, and 1.14. We only formulate it in the nonlattice case.

Theorem 3.23 (i) *Suppose that* $\mu = E[T^{(1)}] < \infty$ *and* $v = E[T^{(2)}] < \infty$. *Then as* $t \to \infty$,

$$\frac{U(t)}{t} \to \frac{1}{\mu} + \frac{1}{v} \quad and \quad U(t+h) - U(t) \to \frac{h}{\mu} + \frac{h}{v}.$$

(ii) *If additionally* $\sigma_1^2 = Var[T^{(1)}] < \infty$ *and* $\sigma_2^2 = Var[T^{(2)}] < \infty$, *then as* $t \to \infty$,

$$U(t) - t\left(\frac{1}{\mu} + \frac{1}{v}\right) \to \frac{R_1(\infty)}{\mu} + \frac{R_2(\infty)}{v},$$

where $R_i(\infty)$ *is given in Theorem 1.14.*

We need the next result to obtain the asymptotic distribution of $N(t)$.

Theorem 3.24 *Suppose* $\mu = E[T^{(1)}], v = E[T^{(2)}], \sigma_1^2 = Var[T^{(1)}], \sigma_2^2 = Var$ $[T^{(2)}]$ *and* $\rho = \rho(T^{(1)}, T^{(2)})$ *are all finite and* $|\rho| < 1$.

(i) *As* $n \to \infty$,

$$\Pr\left\{\frac{S_n^{(1)} - n\mu}{\sigma_1\sqrt{n}} \leq x, \frac{S_n^{(2)} - nv}{\sigma_2\sqrt{n}} \leq y\right\} \to \Pr\{X \leq x, Y \leq y\},$$

where (X, Y) *has the standard bivariate normal distribution with parameter* ρ *(see Remark 3.6).*

(ii) *Suppose that n, m and r are such that $r \to \infty$, $n/r \to K$, and $m/r \to L$, where $0 < K, L < \infty$. Then as $r \to \infty$ we have*

$$\Pr\left\{ \frac{S_n^{(1)} - n\mu}{\sigma_1 \sqrt{n}} \le x, \frac{S_m^{(2)} - m\nu}{\sigma_2 \sqrt{m}} \le y \right\} \to \Pr\left\{ V \le x, W \le y \right\},$$

where (V, W) has the standard bivariate normal distribution with parameter $\rho \min(\sqrt{K/L}, \sqrt{L/K})$.

Proof (i) This is the usual central limit theorem.

(ii) Without loss of generality suppose that $\mu = \nu = 0$ and $\sigma_1 = \sigma_2 = 1$. We consider the case where $K < L$ starting from $n \le m$. We have

$$\left(\frac{S_n^{(1)}}{\sqrt{n}}, \frac{S_m^{(2)}}{\sqrt{m}} \right) = \left(\frac{S_n^{(1)}}{\sqrt{n}}, \frac{S_n^{(2)}}{\sqrt{n}} \frac{\sqrt{n}}{\sqrt{m}} \right) + \left(0, \frac{\sqrt{m - n}}{\sqrt{m}} \frac{1}{\sqrt{m - n}} \sum_{n+1}^{m} Y_i \right),$$

and the two vectors on the right-hand side are independent. Using the central limit theorem (i), we find that

$$\Pr\left\{ \frac{S_n^{(1)}}{\sqrt{n}} \le x, \frac{S_m^{(2)}}{\sqrt{m}} \le y \right\} \to \Pr\{V \le x, W \le y\},$$

where

$$(V, W) = \left(X, Y\sqrt{\frac{K}{L}} \right) + \left(0, Y^\circ \sqrt{\frac{L - K}{L}} \right),$$

(X, Y) has density $\varphi_\rho(x, y)$ and Y° is independent of (X, Y) and has a standard normal distribution. It follows that (V, W) has a standard bivariate normal distribution with parameter $\lambda = \rho \sqrt{K/L}$. The proof follows. $\triangle$

Now we move to the vector $(N^{(1)}(t), N^{(2)}(t))$.

Theorem 3.25 *Suppose that the conditions of Theorem 3.24 hold. For $x, y > 0$ and arbitrary a and b, we have as $t \to \infty$ that*

$$\Pr\left\{ \frac{N^{(1)}(tx) - tx/\mu}{\sqrt{t}} \le a, \frac{N^{(2)}(ty) - ty/\nu}{\sqrt{t}} \le b \right\}$$
$$\to \Pr\left\{ X \le a\frac{\mu^{3/2}}{\sigma_1 \sqrt{x}}, Y \le b\frac{\nu^{3/2}}{\sigma_2 \sqrt{y}} \right\},$$

where (X, Y) has a standard bivariate normal distribution with density $\varphi_\lambda(x, y)$ where $\lambda = \rho \min(x\nu/y\mu, y\mu/x\nu)$.

Proof Recall that

$$\left\{N^{(1)}(t) \le x\right\} = \left\{S_{[x]}^{(1)} \ge t\right\} \text{ and } \left\{N^{(2)}(t) \le y\right\} = \left\{S_{[y]}^{(2)} \ge t\right\}.$$

Define $n = \left[a\sqrt{t} + tx/\mu\right]$ and $m = \left[b\sqrt{t} + ty/v\right]$. As $t \to \infty$, we have $n \sim tx/\mu$, $m \sim ty/v$ and

$$\frac{tx - n\mu}{\sigma_1\sqrt{n}} \to -a\frac{\mu^{3/2}}{\sigma_1\sqrt{x}}, \quad \frac{ty - mv}{\sigma_2}\sqrt{m} \to -b\frac{v^{3/2}}{\sigma_2\sqrt{y}}.$$

Now we have

$$\left\{\frac{N^{(1)}(tx) - tx/\mu}{\sqrt{t}} \le a\right\} = \left\{S_n^{(1)} \ge tx\right\}$$

$$\left\{\frac{N^{(2)}(ty) - tx/v}{\sqrt{t}} \le b\right\} = \left\{S_m^{(2)} \ge ty\right\}.$$

After normalizing, we find that

$$\Pr\left\{S_n^{(1)} \ge tx, S_m^{(2)} \ge ty\right\}$$

$$= \Pr\left\{\frac{S_n^{(1)} - n\mu}{\sigma_1\sqrt{n}} \ge \frac{tx - n\mu}{\sigma_1\sqrt{n}}, \frac{S_m^{(2)} - mv}{\sigma_2\sqrt{m}} \ge \frac{ty - mv}{\sigma_2\sqrt{m}}\right\}.$$

Since $n/t \to x/\mu$ and $m/t \to y/v$, we obtain that

$$\Pr\left\{S_n^{(1)} \ge tx, S_m^{(2)} \ge ty\right\}$$

$$\to \Pr\left\{X \ge -a\frac{\mu^{3/2}}{\sigma_1\sqrt{x}}, Y \ge -b\frac{v^{3/2}}{\sigma_2\sqrt{y}}\right\},$$

where (X, Y) has a standard bivariate normal distribution with density $\varphi_\lambda(x, y)$ where $\lambda = \rho \min(xv/y\mu, y\mu/xv)$. Using the obvious properties of normal distributions, we obtain the desired result. $\triangle$

For the superposed renewal counting process $N(t) = N^{(1)}(t) + N^{(2)}(t)$, we obtain the following corollary

Corollary 3.3 *Under the conditions of the previous proposition, as $t \to \infty$, we have*

$$\Pr\left\{\frac{N(t) - t(1/\mu + 1/v)}{\sqrt{t}} \le x\right\} \to \Pr\{V \le x\},$$

where

$$V = \frac{\sigma_1}{\mu^{3/2}} X + \frac{\sigma_2}{\nu^{3/2}} Y.$$

Proof Taking $x = y = 1$, the previous result gives that

$$\Pr\left\{\left(\frac{N^{(1)}(t) - t/\mu}{\sqrt{t}}, \frac{N^{(2)}(t) - t/\nu}{\sqrt{t}}\right) \le (a, b)\right\}$$

$$\to \Pr\left\{\left(\frac{\sigma_1}{\mu^{3/2}} X, \frac{\sigma_2}{\nu^{3/2}} Y\right) \le (a, b)\right\}.$$

Now we can use Cramer-Wold device to get the result.   △

Remark 3.7 Note that the limiting distribution of V in the Corollary 3.3 is a normal distribution with mean 0 and with variance

$$Var[V] = \frac{\sigma_1^2}{\mu^3} + \frac{\sigma_2^2}{\nu^3} + 2\frac{\sigma_1}{\mu^{3/2}} \frac{\sigma_2}{\nu^{3/2}} \rho \min\left\{\frac{\mu}{\nu}, \frac{\nu}{\mu}\right\}.$$

In the case where the means and the variances are the same, we get

$$Var[V] = 2\frac{\sigma^2}{\mu^3}(1 + \rho).$$

The results in this section are based on the papers of Hunter [21] and Niculescu and Omey [27].

3.5 Applications of the Renewal Processes

3.5.1 Regenerative Processes. Examples

3.5.1.1 Definitions

In Theorem 1.5 we proved that the processes

$$\{M(t) + 1, t \ge 0\} \quad \text{and} \quad \{M(t + T_1), t \ge 0\}$$

have the same finite dimensional distributions.

Consider now a stochastic process $\{X(t), t \ge 0\}$ having the property that *there exist time points at which the process (probabilistically) restarts itself*. More precisely, suppose that with probability 1, there exists a time T_1 such that $\{X(t + T_1)$, $t \ge 0\}$ is a probabilistic replica of the process starting at zero $\{X(t), t \ge 0\}$. This property implies that there exist further times $T_2, T_3, \ldots$ having the same property as T_1. Such a stochastic process is called a *regenerative process*. From the above, it follows that $T_1, T_2, \ldots$ forms a renewal process. We shall say that a cycle is completed every time a renewal occurs.

Remark 3.8 If $\{X(t), t \geq 0\}$ is a regenerative process and $g : \mathbf{R} \to \mathbf{R}$ is a measurable function, then $\{g(X(t)), t \geq 0\}$ is again a regenerative process with the same renewal times.

Remark 3.9 In this section, we consider a few basic results for regenerative processes. For an extensive study of the regenerative processes, we refer to the Serfozo's book [32], Ch. 2.

3.5.1.2 Examples

Example 3.3 A renewal process $\left\{S_0 = 0, S_n = \sum_{i=1}^n T_i, n \geq 1\right\}$ is regenerative and T_1 represents the time of the first regeneration.

Example 3.4 In a discrete recurrent Markov chain T_1 represents the time of the first transition into the initial state and the process is regenerative.

Example 3.5 Consider an inventory model in which the daily demand is represented by r.v.'s $Y_1, Y_2, \ldots$ The Y_i are assumed to be i.i.d. integer valued. The initial inventory is given by $X_1 = S$. The following (s, S) (with $s < S$) ordering policy is used. If the inventory X_n at the beginning of the day n is less than s, then we order enough to bring it back to the initial level S. It is easy to see that $\{X_n, n \geq 1\}$ is a discrete time regenerative process with regeneration point T, the time at which the first order is placed.

3.5.1.3 Ergodic Behavior

Let F denote the d.f. of T and consider $P_A(t) = \Pr\{X(t) \in A\}$ for a Borel set A. By conditioning on T_1, we have

$$
P_A(t) = \int_0^\infty \Pr\{X(t) \in A | T_1 = s\} d F(s)
$$

$$
= \int_0^t \Pr\{X(t) \in A | T_1 = s\} d F(s) + \int_t^\infty \Pr\{X(t) \in A | T_1 = s\} d F(s)
$$

$$
= \int_0^t P_A(t - s) d F(s) + \int_t^\infty \Pr\{X(t) \in A | T_1 = s\} d F(s).
$$

Let us define $q_A(t) = \Pr\{X(t) \in A, T_1 > t\}$. We have

$$
\begin{aligned}
q_A(t) &= \Pr\{X(t) \in A, T_1 > t\} \\
&= \int_0^\infty \Pr\{X(t) \in A, T_1 > t \mid T_1 = s\} dF(s) \\
&= \int_t^\infty \Pr\{X(t) \in A \mid T_1 = s\} dF(s).
\end{aligned}
$$

Going back to the equation above, we have

$$
P_A(t) = q_A(t) + \int_0^t P_A(t - s) dF(s).
$$

This renewal equation can be solved, and we find that

$$
P_A(t) = (U * q_A)(t).
$$

Note that $q_A(t) \leq \Pr\{T_1 > t\}$, and if $\mathbf{E}[T_1] < \infty$, then the integral of $q_A(t)$ is finite. From the Key renewal theorem, we obtain that

$$
P_A(t) \to \frac{1}{\mu} \int_0^\infty q_A(t) dt.
$$

Alternatively, we can write that $q_A(t) = \mathbf{E}\left[\mathbf{I}_{\{X(t) \in A\}} \mathbf{I}_{\{T_1 > t\}}\right]$. This leads to the following interpretation. Let $Y(t)$ be defined as follows:

$$
Y(t) = \begin{cases} 1, & \text{if } X(t) \in A \text{ and } T_1 > t; \\ 0, & \text{otherwise.} \end{cases}
$$

The integral $\int_0^\infty Y(t) dt$ represents the amount of time the process is in A during one cycle. The mean is given by

$$
\mathbf{E}\left[\int_0^\infty Y(t) dt\right] = \int_0^\infty \mathbf{E}[Y(t)] dt = \int_0^\infty q_A(t) dt.
$$

Summarizing, we have proved the following theorem.

Theorem 3.26 *Suppose that* $E[T_1] < \infty$. *Then as* $t \to \infty$, *we have*

$$\Pr\{X(t) \in A\} \to P(A) = \frac{1}{\mu} \int_0^\infty \Pr\{X(t) \in A, T_1 > t\}\,dt.$$

Using the interpretation as above, it follows that

$$\Pr\{X(t) \in A\} \to P(A) = \frac{E[\text{ amount of time in } A \text{ during one cycle}]}{E[\text{ time of one cycle }]}.$$

When $X(t)$ takes values in $\{0, 1, 2, ...\}$, the theorem shows that

$$\Pr\{X(t) = j\} \to P(j) = \frac{1}{\mu} \int_0^\infty \Pr\{X(t) = j, T_1 > t\}\,dt. \tag{3.21}$$

Example 3.6 We consider the inventory example and define $X_1 = S$ and X_n is the inventory at the beginning of day n. The demand on day n is given by Y_n. The order policy is given by:

$$X_{n+1} = \begin{cases} S & \text{if } X_n - Y_n < s; \\ X_n - Y_n & \text{if } X_n - Y_n \geq s. \end{cases}$$

Clearly we have

$$X_1 = S$$
$$X_2 = S - Y_1 \text{ if } S - Y_1 \geq s$$
$$X_3 = S - S_2 \text{ if } S - S_2 \geq s$$
$$\cdots$$
$$X_{n+1} = S - S_n \text{ if } S - S_n \geq s$$
$$X_{n+1} = S \quad \text{if } S - S_n < s.$$

We denote by $S_n = Y_1 + Y_2 + \cdots + Y_n$ the cumulative demand during n consecutive days ($S_0 = 0$). Since Y_n are nonnegative, then the renewal counting process $M(x) = \min\{n : S_n > x\}$, associated with the sequence $S_n, n = 0, 1, 2, \ldots$, is well defined. Denote also $U(x) = E[M(x)]$ the renewal function. Let $T = \min\{n : S - S_n < s\}$. Then the length of one cycle is given by $T = M(S - s)$. It has expected length $E[T] = U(S - s)$. During one cycle we can count the number of times that the inventory is j or more. We count

$$T_j = \min\{n : S - S_n < j\},$$

and find that $T_j = M(S - j)$ and $\mathbf{E}[T_j] = U(S - j), s \leq j \leq S$. From Theorem 3.26, we have

$$\lim_{n \to \infty} \Pr\{X_n \geq j\} = P(j)$$

where $P(j) = 0, j > S, P(j) = 1, j < s$ and

$$P(j) = \frac{U(S - j)}{U(S - s)}, \quad s \leq j \leq S.$$

3.5.1.4 Renewal Rewards

Let us now introduce a reward structure on the regenerative process as follows. When the process is in the set A, we earn a reward given by $f(A)$. Now we have renewal reward process. The next result follows immediately from Theorem 3.18.

Theorem 3.27 *Suppose that* $\mathbf{E}\left| \int\limits_0^{T_1} f(X(t)) \right| dt < \infty$ *and* $\mathbf{E}[T_1] < \infty$. *Then as* $t \to \infty$,

$$\frac{1}{t} \int\limits_0^t f(X(s)) ds \xrightarrow{p} \frac{1}{\mathbf{E}[T_1]} \mathbf{E}\left[\int\limits_0^{T_1} f(X(t)) dt \right],$$

and

$$\frac{1}{t} \mathbf{E}\left[\int\limits_0^t f(X(s)) ds \right] \to \frac{1}{\mathbf{E}[T_1]} \mathbf{E}\left[\int\limits_0^{T_1} f(X(t)) dt \right].$$

If $X(t)$ takes values in $\{0, 1, 2, ...\}$ then under the conditions of Theorem 3.27, one obtains

$$\frac{1}{t} \int\limits_0^t f(X(s)) ds \xrightarrow{p} \sum_{j=0}^{\infty} P(j) f(j),$$

and

$$\frac{1}{t} \mathbf{E}\left[\int\limits_0^t f(X(s)) ds \right] \to \sum_{j=0}^{\infty} P(j) f(j).$$

Taking $f(x) = 1$ if $x = j$ and $f(x) = 0$ otherwise, we have that $\int\limits_0^t f(X(s)) ds$ represents the amount of time that the process is in state j during the time interval $[0, t]$. In this case we have as $t \to \infty$,

$$\frac{\text{amount of time in state } j \text{ during } [0, t]}{t} \xrightarrow{P} P(j),$$

and

$$\frac{\text{mean amount of time in state } j \text{ during } [0, t]}{t} \rightarrow P(j).$$

Example 3.7 Suppose that the lifetime of a car is a r.v. X with d.f. $F(x)$ and that one buys a new car as soon as the car breaks down or reaches S years. Also suppose that a new car costs A and that an additional cost B occurs when the car breaks down before time S. In this example, the time of one cycle is given by $T = \min(X, S)$ and then

$$\Pr\{T \leq t\} = 1 - \Pr\{T > t\} = 1 - \Pr\{X > t, S > t\}.$$

It follows that

$$\Pr\{T \leq t\} = 1, \text{ if } S \leq t, \quad \text{and} \quad \Pr\{T \leq t\} = F(t), \text{ if } S > t.$$

The mean length of one cycle is

$$\mathbf{E}[T] = \int_0^S (1 - F(t))dt.$$

For the costs C, we find that

$$C = A, \text{ if } X > S,$$
$$C = A + B, \text{ if } X \leq S.$$

It follows that

$$\mathbf{E}[C] = A + BF(S).$$

The long run cost per unit time is given by

$$\frac{\mathbf{E}[C]}{\mathbf{E}[T]} = \frac{A + BF(S)}{\int_0^S (1 - F(t))dt}.$$

3.5.1.5 A Queueing Example

Suppose that customers arrive at a single-server service station following a nonlattice renewal process. Upon arrival, a customer is served if the server is idle, and he waits in the queue if the server is busy. The service times are assumed to be i.i.d. and independent of the arrival stream. Let $X_1, X_2, \ldots$ denote the interarrival times of the customers and let $Y_1, Y_2, \ldots$ denote the service times of the customers. Let $n(t)$ denote the number of customers in the system at time t. We suppose that the first customer arrives at the idle system at time 0. Let T_1 denote the next time that a customer arrives and finds the server idle. Then T_1 is a regeneration point for the process $\{n(t), t \geq 0\}$. Now suppose again that the first customer arrives at time 0 at the idle system. If $X_1 > Y_1$, then it follows that at time Y_1 the system is idle and the regeneration cycle is of length X_1. In general, let

$$N = \min \{n : X_1 + X_2 + \cdots + X_n > Y_1 + Y_2 + \cdots + Y_n\},$$

then the busy period will end at $Y_1 + Y_2 + \cdots + Y_n$ and the regeneration cycle will be of length

$$T_1 = X_1 + X_2 + \cdots + X_N.$$

It can be proved that if $E[X] > E[Y]$, then $E[N] < \infty$. Using Wald identity (Lemma 1.2) it follows that $E[T] = \mu E[N]$, where $\mu = E[X]$. Using (3.21), we find that $\Pr\{n(t) = j\} \to P(j)$ exists. Moreover, cf. Theorem 3.27, we have

$$\frac{1}{t} \int_0^t n(s)ds \underset{\to}{P} L = \frac{1}{E[T_1]} E\left[\int_0^{T_1} n(t)dt \right],$$

and

$$\frac{1}{t} E\left[\int_0^t n(s) \right] ds \to L = \frac{1}{E[T_1]} E\left[\int_0^{T_1} n(t)dt \right].$$

Now let W_n denote the waiting time of the nth customer. Since $N = N_1$ is the total number of customers that have been served in one cycle, the sum $A_1 = W_1 + W_n + \cdots + W_{N_1}$ represents the total waiting time of customers in one cycle. By the regeneration property, for the second cycle, the total waiting time is given by $A_2 = W_{N_1+1} + W_{N_1+2} + \cdots + W_{N_1+N_2}$ where N_2 represents the number of customers served in the second cycle. Moreover A_1 and A_2 have the same distribution. In a similar way we find $A_k = W_{N_1+N_2+\cdots+N_{k-1}} + \cdots + W_{N_1+N_2+\cdots+N_k}$. Regarding each A_i as the reward for the ith cycle, by the renewal reward theorem (Theorem 3.18), we find that

$$\frac{1}{n} \sum_{i=1}^n A_i = \frac{1}{n} \sum_{i=1}^{N_1+N_2+\cdots+N_n} W_i \overset{P}{\to} W = \frac{1}{E[N]} E\left[\sum_{i=1}^N W_i \right],$$

and

$$\frac{1}{n} E\left[\sum_{i=1}^{N_1+\cdots+N_n} W_i \right] \to W = \frac{1}{E[N]} E\left[\sum_{i=1}^N W_i \right].$$

To summarize, we have found that

$$L = \frac{1}{E[T]} E \left[\int_0^T n(t)dt \right], \quad W = \frac{1}{E[N]} E \left[\sum_{i=1}^N W_i \right], \quad E[T] = \mu E[N],$$

where L is the average number of customers in the system, W is the average waiting time of a customer, N is the number of customers during one cycle and T is the length of the regeneration cycle. Since (cf. Ross [31], p. 103) we have

$$\sum_{i=1}^N W_i = \int_0^T n(t)dt, \, a.s.,$$

and then it follows that

$$L = \frac{1}{E[T]} E \left[\int_0^T n(t)dt \right] = \frac{1}{E[T]} E \left[\sum_{i=1}^N W_i \right] = W \frac{E[N]}{E[T]}.$$

It follows that $L = \lambda W$, where $\lambda = 1/\mu = $ the arrival rate. This is the law of Little: *the mean number of customers in the system is equal to the mean waiting time of customers in the system multiplied by the arrival rate.*

3.5.2 Optimal Checking in Reliability Theory

When a system fails, it usually fails at a random time T. In many cases, the failure can only be seen by checking the system, otherwise the system keeps on working, but it is functioning in a bad way. Hence, it is useful to check the system from time to time to see whether it has failed or not. It is impossible to check the system too many times, because with each checking of the system, the cost of checking increases. On the other hand, if the system fails at time t_1 and we detect this failure only at time t_2, then a loss occurs during the time interval $t_2 - t_1$. The longer the time between failure and its detection, the greater the loss. This implies that it is useful to check rather frequently. In order to keep the total expected costs under control, it will be necessary to find a balance between the cost of checking and the loss by not detecting the failure.

Example 3.8 ● In the medical world, it is standard that men and women have a regular check at the dentist or at the breast cancer center. For the society, it is useful to know what should be the frequency of such routine health checkings.

- In many countries, drivers have to present their car at a technical center for the first time when the car is 4 years old. From then on they have to present the car every year. From the economic point of view (costs and benefits), perhaps this is not an optimal policy.

In Barlow and Proshan ([4], Theorems 3.2 and 3.3) the authors provide conditions and a procedure to determine times t_1, t_2, ... at which the system should be checked. In Omey [28], the author discussed checking at fixed times $t_1 = b$ and $t_n = b+(n-1)a$, $n \geq 2$. In this section, we discuss the case of checking at *random times*.

Suppose that we have a system that fails at the random time X. The cost of checking the system is a constant c_1 per check. If a failure occurs at time t_1 and we detect this failure at time t_2, then a loss occurs given by $c_2(t_2 - t_1)$. In this section, we propose two policies to check the system at random times.

3.5.2.1 Policy 1

Let $T, T_1, T_2, \ldots$ denote a sequence of i.i.d. random variables and independent of X. Let $\mu = \mathbf{E}[T] < \infty$ and let $S_0 = 0$ and for $n \geq 1$, let $S_n = T_1 + T_2 + \cdots + T_n$. In policy 1, we check the system at times $S_n, n \geq 1$. Let $M(t) = \min \{n \geq 1 : S_n > t\}$ denote the renewal counting process. Given $X = t$, we have to check the system $M(t)$ times in order to detect its failure. Given $X = t$, the total costs are given by

$$C(t \mid X = t) = c_1 M(t) + c_2(S_{M(t)} - t) = c_1 M(t) + c_2 B(t).$$

The expected value of the cost is given by

$$\mathbf{E}[C(t \mid X = t)] = c_1 U(t) + c_2(U(t)\mu - t).$$

In general, we always have $U(t) \geq t/\mu$. From Barlow and Proschan ([4], Theorem 2.5), we have the following result: if $\overline{F}(x) \geq \overline{F}_0(x)$, then $U(t) \leq 1 + t/\mu$. Marshall [24] considered bounds of the form

$$\frac{t}{\mu} + a_l \leq U(t) \leq \frac{t}{\mu} + a_u.$$

Using, for example, the bounds of Barlow and Proschan, we get that

$$c_1 \frac{t}{\mu} \leq \mathbf{E}[C(t \mid X = t)] \leq c_1 \left(1 + \frac{t}{\mu}\right) + c_2 \mu.$$

It follows that

$$c_1 \frac{\mathbf{E}[X]}{\mu} \leq \mathbf{E}[C(X)] \leq c_1 + c_2\mu + c_1 \frac{\mathbf{E}[X]}{\mu}.$$

It is easy to see that the right-hand side in these inequalities is minimized with the choice

$$\mu^* = \sqrt{\frac{c_1 \mathbf{E}[X]}{c_2}}.$$

3.5.2.2 Policy 2

As in the previous section, let $T, T_2, T_3, \ldots$ denote a sequence of i.i.d. random variables and independent of X. Let T_1 denote a r.v. independent of the $T_i, i \geq 2$ and independent of X. Assume that $\mu = \mathbf{E}[T] < \infty$ and $v = \mathbf{E}[T_1] < \infty$. In policy 2 we check the system at times $S_1 = T_1$ and $S_{n+1} = T_1 + T_2 + T_3 + \cdots + T_n, n \geq 1$. Let $M^d(t) = \min \{n \geq 1 : S_n > t\}$ denote the delayed renewal counting process. Recall that

$$\Pr\left\{M^d(t) > n\right\} = \Pr\{S_n \leq t\} = G * F^{(n-1)*}(t), n \geq 1,$$

$$U^d(t) = \mathbf{E}[M^d(t)] = G * \sum_{n=0}^{\infty} F^{n*}(t) = G * U(t).$$

Here F is the d.f. of T and G the d.f. of T_1. Given $X = t$, we have to check the system $M^d(t)$ times in order to detect its failure. Given $X = t$, the total costs are given by

$$C(t \mid X = t) = c_1 M^d(t) + c_2(S_{M^d(t)} - t).$$

Following Wald, we determine $\mathbf{E}[S_{M^d(t)}]$ as follows:

$$\begin{aligned}
S_{M^d(t)} &= T_1 \mathbf{I}_{\{M^d(t)=1\}} + (T_1 + T_2)\mathbf{I}_{\{M^d(t)=2\}} + (T_1 + T_2 + T_3)\mathbf{I}_{\{M^d(t)=3\}} + \cdots \\
&= T_1 \mathbf{I}_{\{M^d(t)\geq 1\}} + T_2 \mathbf{I}_{\{M^d(t)\geq 2\}} + T_3 \mathbf{I}_{\{M^d(t)\geq 3\}} + \cdots \\
&= T_1 + T_2 \mathbf{I}_{\{M^d(t)>1\}} + T_3 \mathbf{I}_{\{M^d(t)>2\}} + \cdots \\
&= T_1 + T_2 \mathbf{I}_{\{T_1 \leq t\}} + T_3 \mathbf{I}_{\{T_1+T_2 \leq t\}} + \cdots
\end{aligned}$$

By independence, it follows that

$$\begin{aligned}
\mathbf{E}[S_{M^d(t)}] &= \mathbf{E}[T_1] + \mathbf{E}[T_2] \Pr\{T_1 \leq t\} + \mathbf{E}[T_3] \Pr\{T_1 + T_2 \leq t\} + \cdots \\
&= v + \mu(G(t) + G * F(t) + G * F^{2*}(t) + \cdots) \\
&= v + \mu U^d(t).
\end{aligned}$$

The expected value of the cost now is given by

$$\mathbf{E}[C(t \mid X = t)] = c_1 U^d(t) + c_2(v + \mu U^d(t) - t).$$

In the special case where we choose $G(x) = F_0(x)$, then $\widehat{U}^d(s) = \widehat{G}(s)/(1 - \widehat{F}(s))$ $= s/\mu$ and we get that $U^d(t) = t/\mu$. It follows that

$$\mathbf{E}[C(t \mid X = t)] = c_1 \frac{t}{\mu} + c_2 \nu.$$

Note that

$$\nu = \lim_{s\downarrow 0} \frac{1 - \widehat{G}(s)}{s} = \lim_{s\downarrow 0} \frac{\mu s - 1 + \widehat{F}(s)}{\mu s^2} = \frac{\mathbf{E}[T^2]}{2\mu}.$$

It follows that

$$
\begin{aligned}
\mathbf{E}[C(X)] &= c_1 \frac{\mathbf{E}[X]}{\mu} + c_2 \frac{\mathbf{E}[T^2]}{2\mu} \\
&= c_1 \frac{\mathbf{E}[X]}{\mu} + c_2 \left(\frac{Var[T]}{2\mu} + \frac{\mu}{2} \right) \\
&= c_1 \frac{\mathbf{E}[X]}{\mu} + c_2 \frac{\mu}{2} + c_2 \frac{Var[T]}{2\mu}.
\end{aligned}
$$

It seems appropriate to choose T in such a way that $Var[T] = 0$. This implies that $T = a$, a constant. Taking $G = F_0$, we get that T_1 has an uniform distribution $T_1 \sim U(0, a)$. We find that

$$\mathbf{E}[C(X)] = c_1 \frac{\mathbf{E}[T]}{a} + c_2 \frac{a}{2}.$$

In order to minimize the expected cost, we take $a^* = \sqrt{2c_1 \mathbf{E}[X]/c_2}$.

3.5.3 Cramer–Lundberg Risk Model

Suppose that the claims arrive at an insurance company at epochs of an ordinary renewal process $\{S_n, n = 1, 2, \ldots\}$ then $N(t), t \geq 0$ is the number of claims arrived in $(0, t]$. Suppose further that the claim amounts are i.i.d. positive random variables $Y_1, Y_2, \ldots$ with d.f. $H(x)$. Assume also that the claim amounts are independent of the renewal process. Then the total claim amount up to time t is

$$Y(t) = \sum_{i=1}^{N(t)} Y_i.$$

In the absence of claims, the reserve of the insurance company increases continuously with intensity c per time unit. If the initial reserve of the company is $x \geq 0$, then the reserve of the company at time t is

$$R(t) = x + ct - Y(t).$$

If the initial reserve of the company is $x \geq 0$, we say that a ruin occurs at time $t \geq 0$ if $Y(t) > x + ct$ and $Y(u) \leq x + cu$ for $u < t$. Let

$$Q(x) = 1 - \Pr\{R(t) < 0, \text{ for some } t \geq 0\}.$$

Then $Q(x)$ is the probability that the ruin will never occur when the initial reserve of the company is x. This model is known as Sparre Andersen risk model. It is introduced by Sparre Andersen [1] as a generalization of the Cramer–Lundberg risk model which assumes that the underlying renewal process $\{S_n\}$, $N(t)$, $t \geq 0$ is a homogeneous Poisson process and so the total claim amount $Y(t)$, $t \geq 0$ by time t is a compound Poisson process.

One of the main problems in these models is the estimation of the probability of ruin $1 - Q(x)$. In this section, we will discuss briefly this problem concerning the Cramer-Lundber risk model. Let us assume that the d.f. of the interarrival times is $F(t) = 1 - e^{-\lambda t}$, $t \geq 0$ so the renewal process is Poisson one with intensity $\lambda > 0$. Assume also that the following basic condition (net profit condition) holds

$$c > \lambda \mathbf{E}[Y], \tag{3.22}$$

that is the income of the company par unit time is greater than the mean claim amount par unit time.

For simplicity we suppose also that $H(0+) = 0$ and $H(x)$ is continuous on $[0, \infty)$. Following Feller ([20], §VI.5) we first derive an integro-differential equation for $Q(x)$. Assume that the first renewal epoch of the Poisson process occurs at time τ and the ruin does not occur at this time, i.e., the first claim amount is $Y_1 = y \leq x + c\tau$. This event has the probability

$$\left(\lambda e^{-\lambda \tau} d\tau\right) \Pr\{Y_1 = y \leq x + c\tau\}.$$

The event that the ruin never occurs after the first claim is independent of the first jump point of the Poisson process and has the probability $Q(x + c\tau - y)$. Then by the total probability formula it follows that

$$Q(x) = \int_0^\infty \int_0^{x+c\tau} Q(x + c\tau - y) dH(y) \lambda e^{-\lambda \tau} d\tau.$$

Substituting $x + c\tau = z$ it follows that

$$Q(x) = \frac{\lambda}{c} \int_x^\infty \left(\int_0^z Q(z - y) dH(y) \right) e^{-\lambda(z-x)/c} dz.$$

This equation shows that $Q(x)$ is differentiable. Differentiating both sides we obtain

$$Q'(x) = \frac{\lambda}{c}\left(Q(x) - \int_0^x Q(x-y)dH(y)\right). \qquad (3.23)$$

Calculating the derivative

$$\frac{d}{dx}\int_0^x Q(x-y)(1-H(y))dy$$

$$= Q(0)(1-H(x)) + \int_0^x Q'(x-y)(1-H(y))dy$$

$$= Q(0)(1-H(x)) - [Q(x-y)(1-H(y))]\,|_{y=0}^{y=x} - \int_0^x Q(x-y)dH(y)$$

$$= Q(x) - \int_0^x Q(x-y)dH(y),$$

we conclude that

$$Q'(x) = \frac{\lambda}{c}\left(\frac{d}{dx}\int_0^x Q(x-y)(1-H(y))dy\right).$$

Integrating both sides on $[0, x]$ one obtains

$$Q(x) = Q(0) + \frac{\lambda}{c}\int_0^x Q(x-y)(1-H(y))dy, \qquad (3.24)$$

which is an equation of renewal type with distribution function $K(x) = \int_0^x (\lambda/c)$
$(1-H(y))dy$. Since $\lim_{x\to\infty} Q(x) = 1$ and $\int_0^\infty (1-H(y))dy = \mathbf{E}[Y] < \infty$ then letting $x \to \infty$ in (3.24) one can apply the dominated convergence theorem to the left-hand side to get

$$1 = Q(0) + \frac{\lambda}{c}\mathbf{E}[Y] \quad \text{or} \quad Q(0) = 1 - \frac{\lambda}{c}\mathbf{E}[Y].$$

Now we are ready to obtain the asymptotic estimation for the ruin probability $1 - Q(x)$. From (3.22) it follows that the distribution function $K(t)$ is improper, i.e., $K(\infty) = \lambda E[Y]/c < 1$. Suppose that there exists a positive constant R such that

$$\int_0^\infty e^{Ry} dK(y) = \frac{\lambda}{c} \int_0^\infty e^{Ry}(1 - H(y))dy = 1, \tag{3.25}$$

which means that the distribution function $K^*(x)$ with density $(\lambda/c)e^{Rx}(1 - H(x))$ is a proper one. Assume also that

$$\mu^* = \int_0^\infty xe^{Rx} dK^*(x) = \frac{\lambda}{c} \int_0^\infty xe^{Rx}(1 - H(x))dx < \infty. \tag{3.26}$$

Theorem 3.28 *If the conditions* (3.25) *and* (3.26) *hold true then*

$$1 - Q(x) \sim \frac{c - \lambda E[Y]}{\mu^* cR} e^{-Rx}, \quad x \to \infty, \tag{3.27}$$

which is known as the Cramer–Lundberg approximation of the ruin probability.

Proof From equation (3.24) we successively obtain

$$Q(x) = 1 - \frac{\lambda E[Y]}{c} + \frac{\lambda}{c} \int_0^x Q(x - y)(1 - H(y))dy,$$

$$1 - Q(x) = \frac{\lambda}{c} \int_x^\infty (1 - H(y))dy + \frac{\lambda}{c} \int_0^x (1 - Q(x - y))(1 - H(y))dy,$$

$$e^{Rx}(1 - Q(x)) = e^{Rx}\frac{\lambda}{c} \int_x^\infty (1 - H(y))dy + \int_0^x e^{R(x-y)}(1 - Q(x - y))e^{Ry}\frac{\lambda}{c}(1 - H(y))dy,$$

$$e^{Rx}(1 - Q(x)) = e^{Rx}\frac{\lambda}{c} \int_x^\infty (1 - H(y))dy + \int_0^x e^{R(x-y)}(1 - Q(x - y))dK^*(y).$$

For the function $z(x) = (\lambda/c)e^{Rx} \int_x^\infty (1 - H(y))dy$ we have that

$$z(x) = (\lambda/c)e^{Rx} \int_x^\infty (1 - H(y))dy \le (\lambda/c) \int_x^\infty e^{Ry}(1 - H(y))dy \downarrow 0, \quad x \to \infty$$

and

$$\int_0^\infty z(x)dx = \frac{\lambda}{c} \int_0^\infty \left(\int_x^\infty e^{Rx}(1 - H(y))dy \right) dx$$

$$= \frac{\lambda}{c} \int_0^\infty \left(\int_0^y e^{Rx}(1 - H(y))dx \right) dy$$

$$= \frac{\lambda}{cR} \int_0^\infty (e^{Ry} - 1)(1 - H(y))dy$$

$$= \frac{\lambda}{cR} \left[\int_0^\infty e^{Ry}(1 - H(y))dy - \int_0^\infty (1 - H(y))dy \right]$$

$$= \frac{\lambda}{cR} \left[\frac{c}{\lambda} - \mathbf{E}[Y] \right].$$

Therefore, $z(x)$ is directly Riemman integrable. Applying the Key renewal theorem (Theorem 1.12) we obtain that

$$\lim_{x \to \infty} e^{Rx}(1 - Q(x)) = \frac{1}{\mu^*} \frac{\lambda}{cR} \left[\frac{c}{\lambda} - \mathbf{E}[Y] \right],$$

which is equivalent to (3.27). $\triangle$

If we additionally assume that the claim amounts are also exponentially distributed with $H(x) = 1 - e^{-\kappa x}$, $\kappa > 0$, then we can find an explicit expression for $Q(x)$. Indeed, in this case the Eq. (3.23) has the form

$$Q'(x) = \frac{\lambda}{c} \left(Q(x) - e^{-\kappa x} \int_0^x Q(y)\kappa e^{\kappa y}dy \right).$$

Differentiating both sides of this equation we get

$$Q''(x) = \frac{\lambda}{c} \left(Q'(x) + \kappa e^{-\kappa x} \int_0^x Q(y)\kappa e^{\kappa y}dy - \kappa Q(x) \right)$$

$$= \frac{\lambda}{c} \left(Q'(x) - \kappa \left(Q(x) - e^{-\kappa x} \int_0^x Q(y)\kappa e^{\kappa y}dy \right) \right)$$

$$= \frac{\lambda}{c} \left(Q'(x) - \kappa \frac{c}{\lambda} Q'(x) \right) = Q'(x) \left(\frac{\lambda}{c} - \kappa \right).$$

The solution of this equation has the form

$$Q(x) = C_1 + C_2 e^{-(\kappa - \lambda/c)x}.$$

Since $E[Y] = 1/\kappa > \lambda/c$ and $Q(\infty) = 1$ it follows that $C_1 = 1$. On the other hand from $Q(0) = 1 - \lambda/(\kappa c)$ it follows that $C_2 = -\lambda/(\kappa c)$. Therefore,

$$Q(x) = 1 - \frac{\lambda}{\kappa c} e^{-(\kappa - \lambda/c)x} \quad \text{or} \quad 1 - Q(x) = \frac{\lambda}{\kappa c} e^{-(\kappa - \lambda/c)x}.$$

A comprehensive treatment of these problems can be found in the books on risk theory (see e.g., Asmussen [3], Grandell [17], Rolski et al. [30]).

3.5.4 Cyclical Ecological-Economic Systems

It is known that ecological-economic systems cycle over time and the dynamics of the system consist of shocks that generate a set of ecological and a set of economic effects. Usually one considers that there are two primary ecosystem functions: exploitation and conservation. Recent research in ecology shows that there are two more functions. One function is that of creative destruction and the other is that of reorganization or renewal, cf. Batabyal [6]. We suppose that shocks to the system occur in accordance with a renewal process with mean interarrival time $E[T] = \mu$. We also suppose that the occurence of a total of S shocks marks the end of a cycle and the beginning of a new cycle. The cost to the society of shocks consist of a fixed cost C_f and a variable cost $c \times s$ that is proportional to the number of shocks. The long run average social cost ASC can be calculated by using the renewal reward theorem. We find that

$$ASC = \frac{E[\text{cycle cost}]}{E[\text{cycle length}]}.$$

The expected length of a cycle is the expected time up to S shocks and it is given by μS. The expected cycle cost is given by

$$E[\text{cycle cost}] = C_f + E[1 \times c \times T_1 + 2 \times c \times T_2 + \cdots + (S - 1) \times c \times T_{S-1}]$$
$$= C_f + c \times \mu \times \frac{S(S - 1)}{2}.$$

It follows that

$$ASC = \frac{C_f}{\mu S} + \frac{c(S - 1)}{2}.$$

It is clear the ASC reaches a minimum for $S = \sqrt{2C_f/\mu c}$.

3.5.5 *Renewal Theory and Environmental Economics*

The design of environmental policy is usually a two step process. According to Cropper and Oates [11] first, standards or targets of environmental quality are set, and second, a regulatory system is designed and put in place to achieve these standards. Here we show how renewal theory can be used to approach the standard setting task in a stochastic cost-benefit framework. Let $\{S_n, n \geq 0\}$ and $\{N(t), t \geq 0\}$ denote the renewal sequence and renewal process. Suppose that the reward earned at the n-th renewal is given by R_n, and let $R(t) = R_1 + R_2 + \cdots + R_{N(t)}$ denote the reward earned by time t. The renewal reward theorem tells us that

$$\lim_{t \to \infty} \frac{E[R(t)]}{t} = \frac{E[R]}{E[T]},$$

where $E[R]$ is the mean reward and $E[T]$ is the mean time between renewals. If we think of a cycle being completed every time a renewal occurs, then the long run expected reward per time unit is given by the expected reward in a cycle divided by the expected length of one cycle. In the context of ecology, we suppose that the state of the system or the quality of the system at time t is represented by $\Sigma(t)$ and we suppose that at time t_0 the system is at level s_0. We also assume that if the quality of the system decreases, then the state of the system $\Sigma(t)$ increases. As the level of pollutants in the system increases, its quality decreases. The goal of the regulators is to set a standard $u > s^\circ$ to ensure that an acceptable level of system quality is maintained. Whenever the system reaches a level $\Sigma(t)$ that exceeds the standard u, the regulators will take action to bring the quality of the system back to an acceptable level s_a. Regulatory actions involve social costs and benefits. If the pollutants of the system hits the level u, then the pollution in the system has to be reduced by an amount of $u - s_a$. Let $C(u, u - s_a)$ and $B(u, u - s_a)$ denote the costs and the (social) benefits of this regulation. The net social benefit is given by

$$NB(u; u - s_a) = B(u, u - s_a) - C(u, u - s_a).$$

When the system gets back to the acceptable level s_a, this action marks the completion of a renewal and the beginning of a new one, (cf. Batabyal and Zoo [5] and Batabyal [7]). We assume that the long-term goal of the regulators is to set the standard u in such a way that the long- term expected net social benefit is maximized. The long-term net social benefit is given by $B_N(u, u - s_a)$ if $\Sigma(t) > u$. The expected value therefore is given by

$$E[R] = NB(u, u - s_a) \Pr\{\Sigma(t) > u \mid \Sigma(t^\circ) = s^\circ\}.$$

Let $E[T] = g(u)$ denote the expected time it takes to get to level u. We find that

$$\frac{E[R]}{E[T]} = \frac{NB(u, u - s_a) \Pr\{\Sigma(t) > u \mid \Sigma(t^\circ) = s^\circ\}}{g(u)}.$$

The best policy corresponds to the value of u that maximizes $\mathbf{E}[R]/\mathbf{E}[T]$. In the special case where $\Sigma(t)$ has a normal distribution $\Sigma(t) \sim N(mt, \sigma^2 t)$, it turns out that $g(u) = (u - s_a)/m$, (cf. Batabyal [7]) and we find that

$$
\begin{aligned}
\frac{\mathbf{E}[R]}{\mathbf{E}[T]} &= \frac{m}{u - s_a} NB(u, u - s_a) \Pr\{\Sigma(t - t^\circ) > u - s^\circ\} \\
&= \frac{m}{u - s_a} NB(u, u - s_a) \Pr\left\{ Z > \frac{u - s^\circ - m(t - t^\circ)}{\sigma\sqrt{t - t^\circ}} \right\},
\end{aligned}
$$

where Z denotes a standard normal random variable. For specific functions $NB(.,.)$ we can maximize this expression with respect to u by using simulation or by using differential calculus.

References

1. Andersen, S.E.: On the collective risk theory in case of contagation between claims. In: Transmisson of XVth International Congress of Acteurs, vol. II, pp. 219–229. New York (1957)
2. Anderson, K.K., Athreya, K.B.: A renewal theorem in the infinite mean case. Ann. Probab. **15**, 388–393 (1987)
3. Asmussen, S.: Ruin Probabilities. World Scientific, Singapore (2000)
4. Barlow, R.E., Proschan, F.: Mathematical Theory of Reliability. Wiley, New York (1965)
5. Batabyal, A.A., Zoo, S.J.: Renewal theory and natural resource regulatory policy under uncertainty. Econ. Lett. **46**, 237–241 (1994)
6. Batabyal, A.A.: Aspects of the optimal management of cyclical ecological-economic systems. Ecol. Econ. **30**, 285–292 (1999)
7. Batabyal, A.A.: A renewal theoretic approach to environmental standard setting. Appl. Math. Lett. **13**, 115–119 (2000)
8. Bickel, P.J., Yahav, J.A.: Renewal theory in the plane. Ann. Math. Stat. **36**, 946–955 (1965)
9. Bingham, N.H., Goldie, C.M., Teugels, J.L.: Regular Variation. Cambridge University Press, Cambridge (1987)
10. Cox, D.R., Smith, W.L.: On the superposition of renewal processes. Biometrika **41**, 91–99 (1954)
11. Cropper, M.L., Oates, W.E.: Environmental economics: a survey. J. Econ. Lit. **30**, 675–740 (1992)
12. Dynkin, E.B.: Some limit theorems for sums of independent random variables with ininite mathematical expectations. Sel. Trans. Math. Stat. Probab. **1**, 171–189 (1961)
13. Erickson, K.B.: Strong renewal theorems with infinite mean. Trans. Am. Math. Soc. **151**, 263–291 (1970)
14. Feller, W.: An Introduction to Probability Theory and its Applications, vol. II. Wiley, New York (1971)
15. Gaigalas, R., Kaj, I.: Convergence of scaled renewal processes and a packet arrival model. Bernoulli **9**, 671–703 (2003)
16. Garsia, A., Lamperti, J.: A discrete renewal theorem with infinite mean. Comment. Math. Helv. **37**, 221–234 (1962)
17. Grandell, J.: Aspects of Risk Theory. Springer, New York (1991)
18. Grigelionis, B.: On a limit theorem in renewal theory. Litovsk Mat. Sb. **2**, 25–34 (1962)
19. Gut, A.: Stopped Random Walks, Limit Theorems and Applications. Springer, New York (1988)
20. Hunter, J.: Renewal theory in two dimensions: basic results. Adv. Appl. Probab. **6**, 376–391 (1974)

21. Hunter, J.: Renewal theory in two dimensions: astymptotic results. Adv. Appl. Probab **6**, 546–562 (1974)
22. Jayakumar, K., Suresh, R.P.: Mittag–Leffler distributions. J. Ind. Soc. Probab. Stat. **7**, 52–71 (2003)
23. Lamperti, J.: An invariance principle in renewal theory. Ann. Math. Stat. **33**, 685–696 (1962)
24. Marshall, K.T.: Linear bounds on the renewal function. SIAM J. Appl. Math. **24**, 245–250 (1973)
25. Mitov, K.V., Yanev, N.M.: Limit theorems for alternating renewal processes in the infinite mean case. Adv. Appl. Probab. **33**, 896–911 (2001)
26. Mitov, K.V., Yanev, N.M.: Superposition of renewal processes with heavy-tailed interarrival times. Stat. Probab. Lett. **76**(6), 555–561 (2006)
27. Niculescu, S., Omey, E.: Asymptotic behavior of the distribution function of multivariate non-homogeneous renewal processes. Study Sci. Math. Hung. **29**, 177–188 (1994)
28. Omey, E.: Optimal checking in reliability theory. In: Proceedings of 8th International Collection on Differential Equations, pp. 347–354. VSP, Plovdiv (1998), (1997)
29. Pillai, R.N.: On Mittag–Leffler functions and related distributions. Ann. Inst. Stat. Math. **42**(1), 157–161 (1990)
30. Rolski, T., Schmidli, H., Schmidt, V., Teugels, J.L.: Stochastic Processes for Insurance and Finance. Wiley, New York (1999)
31. Ross, S.M.: Applied Probability Models with Optimization Applications. Holden-Day, San Francisco (1970)
32. Serfozo, R.: Basic of Applied Stochastic Processes. Springer, New York (2009)

Appendix A
Convolutions and Laplace Transforms

Abstract Certain properties of the convolution and the Laplace transform are collected here for convenience.

A.1 Definitions and Basic Properties

Suppose $g : \mathbf{R} \to \mathbf{R}^+$, $g(t) = 0, t \in (-\infty, 0)$, and $F(t)$ is a distribution function concentrated on $[0, \infty)$. Define the *convolution* of $F(t)$ and $g(t)$ as the function

$$F * g(t) = \int\limits_0^t g(t - u)dF(u), \ t \geq 0,$$

where the integration includes the endpoints. The following properties of convolution hold.

1. $F * g(t) \geq 0, \ t \geq 0.$
2. If $g(t)$ is bounded on finite intervals so is $F * g(t)$.
3. If $g(t)$ is bounded and continuous, then $F * g(t)$ is continuous.
4. The convolution operation can be repeated: $F * (F * g)(t)$.

We denote by

$$I(t) = \begin{cases} 0, \text{ if } t < 0, \\ 1, \text{ if } t \geq 1 \end{cases}$$

the distribution function that assigns a mass 1 at 0.

For any distribution function $F(t)$ concentrated on $\mathbf{R}^+$,

$$F^{0*}(t) = I(t), \ F^{1*}(t) = F(t), \ F^{(n+1)*}(t) = F^{n*} * F(t), \ n \geq 1.$$

K. V. Mitov and E. Omey, *Renewal Processes*, SpringerBriefs in Statistics, DOI: 10.1007/978-3-319-05855-9, © The Author(s) 2014

Clearly, $F^{0*}(t)$ acts as an identity:

$$F^{0*} * g(t) = g(t),$$

and an associative property holds:

$$F * (F * g)(t) = (F * F) * g(t) = F^{2*} * g(t).$$

5. Convolutions of two distribution functions correspond to sums of independent random variables. Let X_1 and X_2 be independent with distribution functions $F_1(t)$ and $F_2(t)$, respectively. Then $X_1 + X_2$ has distribution function $F_1 * F_2(t)$.
6. From 5, it follows that the commutative property holds $F_1 * F_2(t) = F_2 * F_1(t)$.
7. By induction we may show that if $X_1, X_2, \ldots, X_n$ are independent random variables with common distribution function $F(t)$ then $X_1 + \cdots + X_n$ has distribution function $F^{n*}(t)$.
8. If $F_1(t)$ and $F_2(t)$ are absolutely continuous with densities $f_1(t)$ and $f_2(t)$ for $t > 0$ then $F_1 * F_2(t)$ has density for $t > 0$

$$f_1 * f_2(t) = \int_0^t f_1(t - u) f_2(u) du = \int_0^t f_2(t - u) f_1(u) du.$$

For a non-negative random variable X with distribution function $F(t)$, the Laplace transform is a function defined on $[0, \infty)$ by

$$\hat{F}(s) = \mathbf{E}\left[e^{-sX}\right] = \int_0^\infty e^{-st} dF(t), \quad s \geq 0.$$

The following properties are useful:

1. Distinct distributions have distinct Laplace transforms.
2. Suppose X_1, X_2 are independent and have distribution functions $F_1(t)$ and $F_2(t)$, respectively. Then

$$\widehat{(F_1 * F_2)}(s) = \hat{F}_1(s)\hat{F}_2(s).$$

3. If $F(t)$ is a distribution function, then

$$\widehat{F^{n*}}(s) = (\hat{F}(s))^n.$$

4. For $s > 0$, $\hat{F}(s)$ has derivatives of all orders, and for any $n \geq 1$,

$$(-1)^n \frac{d^n}{ds^n} \hat{F}(s) = \int_0^\infty e^{-st} t^n dF(t).$$

Now, by monotone convergence,

$$\lim_{s \downarrow 0} (-1)^n \frac{d^n}{ds^n} \hat{F}(s) = \int_0^\infty t^n dF(t) \leq \infty.$$

In particular, if the random variable X has $F(t)$ as its distribution function, then $\mathbf{E}[X] = -\hat{F}'(0)$ and $\mathbf{E}[X^2] = \hat{F}''(0)$, and so on.

5. An integration by parts proves the following formulas

$$\int_0^\infty e^{-st} F(t)dt = \frac{\hat{F}(s)}{s}, \quad \int_0^\infty e^{-st}(1 - F(t))dt = \frac{1 - \hat{F}(s)}{s}. \tag{A.1}$$

We now extend these notions to arbitrary distributions and measures U on $[0, \infty)$. Suppose that U is a measure on $\mathbf{R}^+$. Then $U(t) = U([0, t])$ is a non-negative and nondecreasing function on $[0, \infty)$, but perhaps, $U(\infty) = U([0, \infty)) = \lim_{t \uparrow \infty} U(t) > 1$.

If there exists $a \geq 0$ such that

$$\int_0^\infty e^{-st} dU(t) < \infty$$

for $s > a$ then

$$\hat{U}(s) = \int_0^\infty e^{-st} dU(t) < \infty, \ s > a \tag{A.2}$$

is called the Laplace transform of $U(t)$. If such a does not exist, we say the Laplace transform is undefined. For detailed discussions on these topics we refer to the book of Feller [2].

A.2 Regularly Varying Functions and Tauberian Theorems

We need the following definitions.

Definition 1.1 A measurable function $L : [A, \infty) \to \mathbf{R}^+$, where $A \geq 0$, is said to be slowly varying at infinity (s.v.f.) if for every $x > 0$,

$$\lim_{t \to \infty} \frac{L(tx)}{L(t)} = 1. \tag{A.3}$$

Definition 1.2 A measurable function $f : \mathbf{R}^+ \to \mathbf{R}^+$ is said to be regularly varying at infinity (r.v.f.) with exponent $\rho > 0 (f(t) \in RV(\rho))$ if

$$f(t) \sim t^\rho L(t), \quad t \to \infty, \tag{A.4}$$

for some s.v.f. $L(t)$.

Below, we formulate only the results for regularly varying functions that we need in the book. For a comprehensive treatment of the notion of regular variation and its applications, we refer to the books of Feller [2], Bingham et al. [1], or Seneta [3].

Theorem 1.1 (Feller [2], Theorem 1, §VIII.9) (a) *If $Z(t) \in RV(\gamma)$ and the integral*
$$Z_p^*(x) = \int_x^\infty t^p Z(t) dt \text{ converges then}$$

$$\frac{t^{p+1} Z(t)}{Z_p^*(t)} \to \zeta,$$

where $\zeta = -(p + 1 + \gamma) \geq 0$. Conversely, if the last relation holds true and $\zeta > 0$ then Z and Z_p^* are regularly varying with exponents $\gamma = -\zeta - p - 1$ and $-\zeta$, respectively.

(b) *If $Z \in RV(\gamma)$ and the integral $Z_p(x) = \int_0^x t^p Z(t) dt$ converges then if $p \geq -\gamma - 1$, then*
$$\frac{t^{p+1} Z(t)}{Z_p(t)} \to \zeta,$$

where $\zeta = p + \gamma + 1$. Conversely, if the last limit is $\zeta > 0$ then Z and Z_p are regularly varying with exponents $\zeta - p - 1$ and ζ, respectively.

Theorem 1.2 (Feller [2], Lemma, §XIII.5) *If*

$$U(t) \sim \frac{t^\rho L(t)}{\Gamma(\rho + 1)}, \quad t \to \infty,$$

for $\rho > 0$ and has $u(t) = U'(t)$ is eventually monotone then

$$u(t) \sim \frac{\rho U(t)}{t}, t \to \infty.$$

Theorem 1.3 *Let $U(t)$ be a measure on $\mathbf{R}^+$ and $\hat{U}(s)$ be its Laplace transform defined by (A.2). Then*

$$U(\infty) := \lim_{t \to \infty} U(t) < \infty \text{ if and only if } \hat{U}(0) := \lim_{s \to 0} \hat{U}(s) < \infty. \tag{A.5}$$

If this is the case then $U(\infty) = \hat{U}(0)$.

Theorem 1.4 (Karamata's Tauberian theorem) *If $L(t)$ is slowly varying at infinity and $0 \le \rho < \infty$, then each of the relations*

$$\hat{U}(s) \sim s^{-\rho} L\left(\frac{1}{s}\right), \quad s \to 0, \tag{A.6}$$

and

$$U(t) \sim \frac{1}{\Gamma(\rho + 1)} t^{\rho} L(t), \quad t \to \infty, \tag{A.7}$$

implies the other.

The proofs and certain comments can be found in the books cited above.

The next two theorems describe the limiting behavior of the sum of independent identically distributed random variables in case when their mathematical expectation is infinite.

Theorem 1.5 (Feller [2], Theorem 1, §XIII.6) *Let $\beta \in (0, 1)$ be fixed. The function $\gamma_\beta(s) = e^{s^\beta}$, $s \ge 0$ is the Laplace transform of a distribution function $G_\beta(t)$, which has the following properties:*

(a) *$G_\beta(t)$ is stable, that is, if $X_1, X_2, \ldots, X_n$ are random variables with distribution function G_β then*

$$\frac{X_1 + X_2 + \cdots + X_n}{n^{1/\beta}}$$

has the same distribution function $G_\beta(t)$.

(b) *$G_\beta(t)$ satisfies the relations:*

$$t^\beta (1 - G_\beta(t)) \to \frac{1}{\Gamma(1 - \beta)}, \quad t \to \infty$$

$$\exp(t^{-\beta}) G_\beta(t) \to 0, \quad t \to \infty.$$

Theorem 1.6 (Feller [2], Theorem 2, §XIII.6) *Let $F(t)$ be a distribution function on $(0, \infty)$, i.e., $(F(0) = 0, F(+\infty) = 1)$ and such that*

$$F^{*n}(a_n t) \to G(t), \tag{A.8}$$

for the points of continuity of G, where G is a proper distribution function, not concentrated in one point. Then

(a) *There exists a slowly varying at infinity function L and a constant β, $0 < \beta < 1$, such that*

$$1 - F(t) \sim \frac{t^{-\beta} L(t)}{\Gamma(1 - \beta)}, \quad t \to \infty.$$

(b) *Conversely, if F satisfies the last relation then it is possible to choose a sequence a_n, $n = 1, 2, \ldots$, such that*

$$\frac{nL(a_n)}{a_n^\beta} \to 1,$$

and in this case (A.8) holds with $G(t) = G_\beta(t)$.

Remark 1.1 If $X_1, X_2, \ldots$ are independent identically distributed random variables with distribution function $F(t)$ then (A.8) means that

$$\lim_{n\to\infty} \Pr\left\{\frac{X_1 + X_2 + \cdots + X_n}{a_n} \le t\right\} = G(t).$$

This is an analog of the central limit theorem in the case of a sum of i.i.d. random variables with infinite mean.

References

1. Bingham, N.H., Goldie, C.M., Teugels, J.L.: Regular Variation. Cambridge University Press, Cambridge (1987)
2. Feller, W.: An Introduction to Probability Theory and its Applications, vol. II. Wiley, New York (1971)
3. Seneta, E.: Regularly Varying Functions. Springer, Berlin (1976)